Rebecca Basile

Thermoregulation and Resource Management in Honeybees (Apis mellifera)

Rebecca Basile

Thermoregulation and Resource Management in Honeybees (Apis mellifera)

The Distribution of Resources via Trophallaxis and its Impact on the Heating Performance in Western Honeybee Colonies

Südwestdeutscher Verlag für Hochschulschriften

Impressum/Imprint (nur für Deutschland/ only for Germany)
Bibliografische Information der Deutschen Nationalbibliothek: Die Deutsche Nationalbibliothek verzeichnet diese Publikation in der Deutschen Nationalbibliografie; detaillierte bibliografische Daten sind im Internet über http://dnb.d-nb.de abrufbar.
 Alle in diesem Buch genannten Marken und Produktnamen unterliegen warenzeichen-, markenoder patentrechtlichem Schutz bzw. sind Warenzeichen oder eingetragene Warenzeichen der jeweiligen Inhaber. Die Wiedergabe von Marken, Produktnamen, Gebrauchsnamen, Handelsnamen, Warenbezeichnungen u.s.w. in diesem Werk berechtigt auch ohne besondere Kennzeichnung nicht zu der Annahme, dass solche Namen im Sinne der Warenzeichen- und Markenschutzgesetzgebung als frei zu betrachten wären und daher von jedermann benutzt werden dürften.

Verlag: Südwestdeutscher Verlag für Hochschulschriften Aktiengesellschaft & Co. KG
Dudweiler Landstr. 99, 66123 Saarbrücken, Deutschland
Telefon +49 681 37 20 271-1, Telefax +49 681 37 20 271-0
Email: info@svh-verlag.de
Zugl.: Würzburg, Universität, Diss., 2009

Herstellung in Deutschland:
Schaltungsdienst Lange o.H.G., Berlin
Books on Demand GmbH, Norderstedt
Reha GmbH, Saarbrücken
Amazon Distribution GmbH, Leipzig
ISBN: 978-3-8381-1308-1

Imprint (only for USA, GB)
Bibliographic information published by the Deutsche Nationalbibliothek: The Deutsche Nationalbibliothek lists this publication in the Deutsche Nationalbibliografie; detailed bibliographic data are available in the Internet at http://dnb.d-nb.de.
 Any brand names and product names mentioned in this book are subject to trademark, brand or patent protection and are trademarks or registered trademarks of their respective holders. The use of brand names, product names, common names, trade names, product descriptions etc. even without a particular marking in this works is in no way to be construed to mean that such names may be regarded as unrestricted in respect of trademark and brand protection legislation and could thus be used by anyone.

Publisher: Südwestdeutscher Verlag für Hochschulschriften Aktiengesellschaft & Co. KG
Dudweiler Landstr. 99, 66123 Saarbrücken, Germany
Phone +49 681 37 20 271-1, Fax +49 681 37 20 271-0
Email: info@svh-verlag.de

Printed in the U.S.A.
Printed in the U.K. by (see last page)
ISBN: 978-3-8381-1308-1

Copyright © 2010 by the author and Südwestdeutscher Verlag für Hochschulschriften Aktiengesellschaft & Co. KG and licensors
All rights reserved. Saarbrücken 2010

Against Idleness And Mischief

How doth the little busy bee
Improve each shining hour,
And gather honey all the day
From every opening flower!

How skilfully she builds her cell!
How neat she spreads the wax!
And labours hard to store it well
With the sweet food she makes.

In works of labour or of skill,
I would be busy too;
For Satan finds some mischief still
For idle hands to do.

In books, or work, or healthful play,
Let my first years be passed,
That I may give for every day
Some good account at last.

Isaac Watts (1674-1748)

The Crocodile

How doth the little crocodile
Improve his shining tail,
And pour the waters of the Nile
On every golden scale!

How cheerfully he seems to grin!
How neatly spread his claws,
And welcomes little fishes in
With gently smiling jaws!

Lewis Carroll (1832 – 1898)

Table of contents

1. Introduction   **7**
    1.1   Subject of the dissertation   7
    1.2   Evolution of Eusociality   8
       1.2.1   Altruism, kin selection, Hamilton´s rule and the odds against altruism   8
       1.2.2   Social interaction, cooperation and dominance hierarchies in animal groups   12
    1.3   The Western honeybee – *Apis mellifera*   14
       1.3.1   Natural range and characteristics   14
       1.3.2   Colony structure   16
       1.3.3   Division of labor, task allocation and life span in the honeybee   17
          1.3.3.1   Summer bees   18
          1.3.3.2   Winter bees   19
       1.3.4   Juvenile hormone (JH)   20
       1.3.5   Vitellogenin   20
       1.3.6   Genetic influence on the division of labor   22
       1.3.7   The nest of the honeybee   23
       1.3.8   Thermoregulation and heating activity   24
       1.3.9   Trophallaxis   30
       1.3.10   The morphology of the antenna   35
    1.4   Specific aim   37

2. Antennal dexterity in honeybees – about the lopsided use of the antennae in trophallactic contacts   **40**
    2.1   Abstract   40
    2.2   Introduction   40
    2.3   Materials and Methods   42
    2.4   Results   44
    2.5   Discussion   46
    2.6   Appendix – Figures and Tables   50

3. Does sugar equal heat? – Sugar intake and its impact on thoracic heat production in the honeybee   **58**
    3.1   Abstract   58

3.2	Introduction	59
3.3	Materials and Methods	63
	3.3.1 Set up without additional water	64
	3.3.2 Set up with additional water	65
3.4	Results	65
	3.4.1 Set up without additional water	65
	3.4.2 Set up with additional water	66
3.5	Discussion	67
3.6	Appendix – Figures and Tables	75

4. Trophallactic activities in the brood nest – heaters get supplied with high performance fuel — **87**

4.1	Abstract	87
4.2	Introduction	88
4.3	Materials and Methods	91
	4.3.1 Behavioral observations	91
	4.3.2 Thermal imaging	92
4.4	Results	93
	4.4.1 Behavior of donors and recipients	93
	4.4.2 Thoracic temperature and trophallaxis	95
4.5	Discussion	96
4.6	Appendix – Figures and Tables	100

5. Heat seeker – The honeybee feeding activity has a thermal trigger — **106**

5.1	Abstract	106
5.2	Introduction	106
5.3	Materials and Methods	109
	5.3.1 Behavioral observations in the hive	109
	5.3.2 Warm-up experiments in the hive	111
	5.3.3 Warm-up experiments in the arena	112
	5.3.4 Warm-up experiments with restrained bees	112
5.4	Results	113
	5.4.1 Trophallactic behavior in the observation hive	113
	5.4.2 Feeding contacts	114
	5.4.3 Winter experiments with artificial heating	114

5.4.4	Arena experiments	115
5.4.5	Reactions to heat and electromagnetic fields	115
5.5	Discussion	116
5.6	Appendix – Figures and Tables	123

6. General discussion   134

7. Summary   143

8. Index of Figures   145

9. Index of Tables   148

10. Index of Abbreviations   150

11. References   153

Introduction

1. Introduction

1.1 Subject of the dissertation

The ecological success of social insects is largely based on the complex organization of their colonies. Even though there is no control that coordinates their action, the members of a colony specialize on certain of the various tasks by division of labor (OSTER & WILSON, 1978; BOURKE & FRANKS, 1995).

In a honeybee colony the adjustment of the labor devoted to tasks inside and outside the hive is expected to be highly adaptive. Biotic and abiotic factors like temperature, brood rearing conditions, pollen and nectar availability strongly fluctuate and therefore condition which tasks have priority and require increased attention.

The division of labor between the members of the hive is implemented by temporal polytheism in which the worker's physiological state and its probability of task performance correlate with its age. Specializations are therefore temporary (RÖSCH, 1927).

Physical polytheism, as it occurs in many ants and termites, does not occur in honeybees (WILSON, 1971; OSTER & WILSON, 1978). Nevertheless, there are differences between the individuals in one colony, concerning their task performances. There is evidence for lifetime differences in behavioral preferences which cannot be explained by differences in adult development. Some tasks like guarding or undertaker duties are only performed by a small percentage of a colony's workers. In this context several studies showed that due to the genetic variance in the colony different tasks are accomplished with more constancy than in a hive with higher genetic relatedness (ROBINSON, 1992).

Beside a genetic basis of the division of labor other physiological factors seem to influence task related behaviors, like the levels of vitellogenin and juvenile hormone which are related to behavioral development in adult honeybees (ROBINSON ET AL., 1989; FAHRBACH & ROBINSON, 1996; HUANG & ROBINSON, 1996).

Most of these studies concentrated on behavioral tasks that are related to communication, reproduction, foraging behavior and recruitment. In social insects relatively little is known about how the built-up stocks are organized and distributed within a colony consisting of several thousands of individuals.

Introduction

The central questions of this thesis are how these resources are shared; whether there is a performance-related reward system; what regulates individual difference in performance and how such systems might have evolved in the context of task allocation, division of labor and the energy balance in the lives of individuals and the colony.

The various methods used for fielding these questions are adapted to the respective field of research science. Methods from classical behavioral ecology, behavioral physiology, neurobiology, and theoretical approaches were used as tools in this thesis.

1.2 Evolution of Eusociality

The practice of individuals or larger societal entities working together instead of working separately in competition is known in many animal groups. The benefits animals achieve by hunting, foraging or defending collectively are obvious. Why should there be a distinction between cooperation among unrelated individuals and cooperation among related individuals?

If the individual benefit is measured in successful reproduction, the relatedness between cooperating individuals gains in importance.

1.2.1 Altruism, kin selection, Hamilton´s rule and the odds against altruism

A social behavior of sacrificing one's own reproductive potential to benefit another individual is called altruism. Altruistic behavior, especially the reproductive division of labor, is opposed to the fundamental idea of natural selection and can only be explained by a complex system of indirect individual fitness gain. This individual-level or gene-level selection states that each member of the colony has been selected to maximize its own reproductive success (inclusive fitness). Group behaviors such as cooperative food collection, the defense of the hive, the feeding of the brood, and thermoregulation are simply statistical summations of many individuals' ultimately selfish actions (HAMILTON, 1964, 1972; DAWKINS, 1976, 1982).

Hamilton discriminates between "direct fitness", concerning genes that can be passed to the next generation directly by the individual trough reproduction, and "indirect fitness", referring to genes that are passed to the next generation by helping the reproductive success of kin.

Introduction

An ordinary diploid, sexually produced organism shares 50 % of its genes with either of his parents. Accordingly, it shares about 50 % with its siblings, 25 % with its uncles, aunts, grandparents, grandchildren and so forth. Hamilton's stroke of genius was to reformulate the definition of fitness as the number of an individual's alleles in the next generation. Or, more precisely, inclusive fitness is defined as an individual's relative genetic representation in the gene pool of the next generation (Fig. 1.1).

Under certain circumstances, altruistic behavior towards kin (indirect fitness) can enhance the inclusive fitness dramatically. This interrelation is abstracted as Hamilton's rule:

Altruism will occur when:

"**c** *(cost to the individual)* is **lower** than **(r) b** *(benefit to the kin)*"

An explanation for the altruistic behavior on one hand and the fundamental genetically egoism in honeybees on the other hand can be found in the kin selection theory.

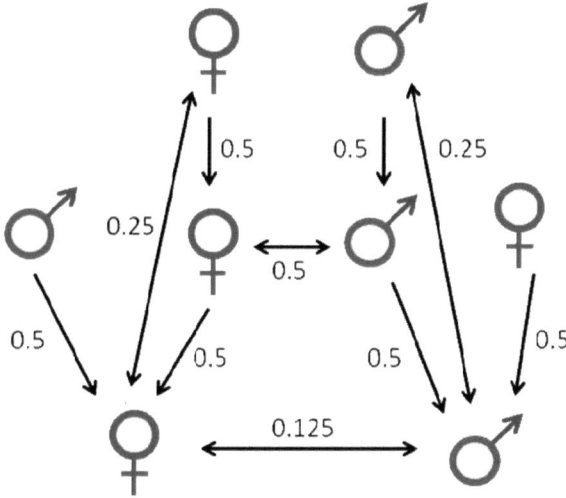

Fig. 1.1 Coefficient of relatedness in diploid organisms

Every parent (top row) transmits 50 % of its genetic information to each offspring (middle row). On the average, siblings therefore share half of each parent's contribution to their genome, adding to a coefficient of relatedness $r=0.5$. Consequently, cousins share an $r=0.125$ or $r=1/8$ (bottom row). Likewise, these cousins are related to their common grandparents by ¼ or $r=0.25$. One may also say that r is a measure for the probability that any given allele is shared by two individuals.

Introduction

Kin selection refers to changes in gene frequency across generations which are driven at least in part by interactions between related individuals, and this forms much of the conceptual basis of the theory of social evolution (HAMILTON, 1964). A gene that prompts behavior that enhances the fitness of relatives but lowers that of the individual displaying the behavior, may nonetheless increase in frequency, because relatives often carry the same gene; this is the fundamental principle behind the theory of kin selection. According to the theory, the enhanced fitness of relatives can at times more than compensate for the fitness loss incurred by the individuals displaying the behavior (inclusive fitness) (QUELLER & STRASSMAN, 2002; WEST ET AL. 2006).

While most animal genera have a hetero- and a homogametic sex, hymenopterans universally produce males from unfertilized, haploid eggs and females from fertilized, diploid eggs. This system skews relatedness in an almost perfect way for eusociality to evolve. A female worker's genome comes half from the father (haploid) and half from the mother (diploid). That means she carries all of her father's genes and half of her mother's genes. So does her sister, implying that they share of course the entire genome of their common father, plus, on average, a quarter of their mother's genome, yielding a coefficient of relatedness of 0.75 (Fig. 1.2). Therefore, altruistically helping their mother and her offspring needs only to yield a small benefit compared to "normal" diploid organism in order to spread through the population. So much for the theory.

In reality, although all members of a honeybee colony usually share the same mother, the female members do not share the same father, a fact of major importance in understanding the evolution of honeybee social life. Mating by queen honeybees occurs only during the two-week period immediately following the queens' emergences as adults. During this time, a queen makes several flights from her nest, receiving sperm from 5 to 30 different drones (ADAMS ET AL., 1977; ESTOUP ET AL., 1994; FUCHS & MORITZ, 1999; NEUMANN ET AL., 1999, NEUMANN & MORITZ, 2000) on one to four of these flights. By the close of her mating period, each queen has stored about 5 million spermatozoae in her spermatheca, a sufficient supply for her potential lifespan of roughly three years (ROBERTS, 1944; WOYKE, 1962; 1964).

DAWKINS (1976) acknowledges this difficulty at the conclusion of his explanation of this spectacular triumph of sociobiology, but the response he offers is: *"My head is now spinning, and it is high time to bring this topic to a close."*

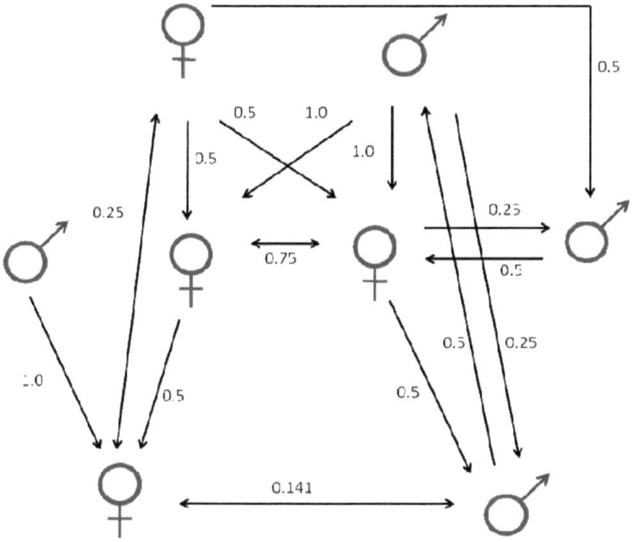

Introduction

Fig. 1.2 Coefficient of relatedness with haplo-diploid sex determination

The coefficients are skewed with respect to the diploid system depicted in Fig. 1.1 For example, sisters (middle row) are more related to each other (r=0.75) than they are to their mother (top row; r=0.5).

From a biological point of view, altruism should not exist. The Darwinian theory of natural selection holds that those organisms survive and reproduce which are best adapted to their environment. They are "selected" by the natural processes of geography, climate, food supplies, predation, etc. To that extent, any organism that devotes itself to the welfare of other organisms jeopardizes its own prospects for reproduction and enhances those of the recipient of the assistance. As that trend continues, the altruist strain would seem bound to be selected out of existence. This line of reasoning has been backed by the development of the more precise investigations of genetics in this century.

Additional difficulties in explaining kin selection and altruism in honeybees arise because workers never actually try to rear their own offspring as long as they can help their mother. Apart from the physical incapability to mate and lay fertilized eggs, an important limit to a worker's success in personal reproduction would undoubtedly be her ability to care for her offspring, for example feeding it and keeping it warm at the same time. If the worker attempted to go it completely alone, she would face the many hurdles of solitary life, including constructing a nest, laying eggs, and feeding and guarding her brood. Thus it seems likely that the cost of altruism by workers is negligible (SEELEY, 1985)

On the other hand, there are situations like swarming, the death of the old and the raising of a new queen or the drifting of honeybee workers. In these situations the relatedness of workers and queen (and accordingly her offspring) declines. In particular cases, the relatedness between worker and queen is 0. Nevertheless, all workers perform their tasks and the colony continues to function as a superorganism, no matter how low the relatedness towards queen and offspring gets. Swarming is in fact an annual event, and therefore naturally lowers the relatedness in the newly established colony for the first generations.

1.2.2 Social interaction, cooperation and dominance hierarchies in animal groups

Animal populations are often organised into groups. These groups differ in characteristics such as composition, size, permanency, coordination, cohesion, and social formation (HEMELRIJK, 2002). A group may form for simple purposes such as feeding, drinking, or mating. In contrast, a true animal society is a remarkable group of individuals of the same species that maintain a cooperative social relationship.

A society of animals usually has some maintenance of social structure and spacing of group members. A colony of social insects consisting of tens of thousands of individuals is able to cope with huge socio-economic demands like foraging, building and cleaning the nest, and nursing brood.

Group members are able to divide the work efficiently among them. Such a division of labor is flexible, i.e. the ratio of workers performing different tasks varies according to the changing needs and circumstances of the colony.

This task division may be based on different mechanisms, like a genetic difference in predisposition (ROBINSON & PAGE, 1988; ROBINSON & PAGE, 1989; MORITZ ET AL., 1996), or the response-threshold to perform certain tasks. These mechanisms may be combined with a self-reinforcing learning process (THERAULAZ ET AL., 1998).

The division of tasks may also be a consequence of dominance relations. An agonistic behaviour, in which one animal is aggressive or attacks another animal, which responds either by returning the aggression or submitting, is often responsible for the patterns that

Introduction

account for dominance relations. This agonistic behavior has generally become known as the "pecking order", which was described first by SCHJELDERUP-EBBE (1922) in chickens.

Social dominance has been considered to be of fundamental social importance (GARTLAN, 1968), but this explanatory value was challenged. Central to the debate is the relationship between dominance and aggression (FRANCIS, 1988). There are two opposing views. On one hand a higher rank is believed to offer optimal access to resources, and therefore individuals should seize every opportunity to increase their rank (POPP & DEVORE, 1979). On the other hand, the function of a dominance hierarchy is thought to reduce costs associated with aggression, and therefore, individuals should avoid conflict as soon as relationships are clear.

Such relationships have been described for bumblebees (*Bombus terrestris*) by VAN HONK and HOGEWEG (1981) and HOGEWEG and HESPER (1983, 1985).

In an experimental study VAN HONK and HOGEWEG (1981) discovered that during the growth of the colony workers developed into two types, the low-ranking so called "common", and the high-ranking, so-called "elite" workers.

The behavioural patterns of these two types of workers differ noticeably: whereas the "common" workers mainly forage and take rest, the elite workers are more active, feed the brood, interact frequently with each other and with the queen, and sometimes lay eggs.

In order to study the minimal conditions needed for the formation of the two types of workers, HOGEWEG and HESPER (1983, 1985) set up a so-called "Mirror" model. It contains biological data concerning time and development of the offspring. Space parameters are reflected in the peripheral areas (where the commons work) and areas, where the elite works. The artificial bumblebees operate locally insofar as their behavior is triggered by what they encounter. For example, if an adult bumblebee meets a larva, it feeds it.

When an adult meets another, a dominance interaction takes place, the outcome of which (dominant or submissive behavior) is self-reinforcing. All workers start with the same dominance rank. This model automatically generates two stable classes, those of "commons" (low-ranking workers) and "elites" (high ranking workers) with their typical conduct. This difference occurs only if the nest is separated into a center and a periphery, as it is found in real nests. If the work force is cut in half, encounters between workers and brood become more frequent, whereas encounters between workers and other workers decrease in frequency. Therefore, the distribution of work shifts towards the "commons".

Introduction

The foraging and brood feeding activity has to be upheld by fewer workers, which is achieved without differentiating the workers by their individual threshold level or their genes, just by following simple rules of dominance hierarchy in a group (HEMELRIJK, 2002).

Similar behavior is documented in the eusocial wasp *Polybia occidentalis* by O'DONNELL (2001). *Polybia* displays a behavior described as "social biting". This ritualized aggressive behavior influences foraging rates. Bitten wasps left the nest to go foraging, while the biting wasps stay in the nest, respectively on the nest surface. O'DONNELL gathered from his observations in *P. occidentalis* and in other eusocial insects with large worker forces, that biting and other types of social contact among workers may regulate task performance independently of direct reproductive competition (O'DONNELL, 2001).

As already mentioned, in dominant hierarchies a group of animals is organized in a way that offers some members of the group greater access to resources, such as food or mates, than others. High dominance rank in a group is supposed to be associated with benefits such as easier access to mates, food and safe spatial location. The safest location is in the center of the group, because there individuals are protected by other group-members who shield them from atmospheric exposure and predators approaching from outside. Therefore, according to the well-known "selfish herd"-theory of HAMILTON (1971), individuals have evolved a preference for a position in the center, the so-called "centripetal instinct". The competition for a position in the center is won by dominants, and thus, dominants will end up in the center. This is thought to be the main reason why in many animal species dominants are seen to occupy the center of a group (HEMELRIJK, 2002).

1.3 The Western honeybee – *Apis mellifera*

1.3.1. Natural range and characteristics

The Western honeybee is one of nine extant and living species of *Apis* (JOHANNESMEIER, 2001). This genus is native to Europe, Africa, and Asia (RUTTNER, 1988).

Honeybees also thrive in North America, South America and Australia, but only since European man introduced them at various times during the seventeenth to nineteenth centuries in course of the colonization and the large-scale emigration of these periods (CRANE, 1999)

Introduction

The western honeybee's natural distribution extends from the steppes of western Asia through Europe as far north as southern Norway and into all of Africa, except its great desert areas.

The most common European subspecies of the Western honeybee are *A. mellifera carnica*, *A. m. mellifera* (northern Europe), *A. m. ligustica* (Italy), *A. m. caucasica* (Caucasus) and *A. m. macedonica* (southeastern Europe). Subspecies like *A. m. cecropia*, *A. m. sicula* and *A. m. iberica* are considered as highly endangered or even as extinct (RUTTNER, 1988) (Fig. 1.3). Even the once widely distributed subspecies *A. m. mellifera* seems to be endangered mainly via hybridization with other subspecies (SOLAND-RECKEWEG ET AL., 2008).

Fig. 1.3 European subspecies of *Apis mellifera* and their distribution

The most common African subspecies are *A. m. intermissa* (northern Africa), *A. m. scutellata* (southern and central Africa) and *A. m. capensis* (HEPBURN & RADLOFF, 1998).

The Carniolan honeybee differentiated together with the Italian bee, *A. m. ligustica*, during the last ice age, when honeybees in Europe existed as isolated populations in a few southerly refugia (WHITFIELD ET AL., 2006). The Italian bee is adapted to the short and mild Mediterranean winters.

Experimental studies have revealed that Carniolan bees range farther in foraging, choose larger nest capacities and disperse farther from the nest when reproducing their colonies than *A. m. mellifera* (BOCH, 1957; JAYCOX & PARISE, 1980, 1981; GOULD, 1982).

1.3.2 Colony structure

Unlike other social hymenoptera the honeybees' colonies are perennial and potentially "immortal" (TAUTZ & HEILMANN, 2007). The population of a honeybee colony in a good summer can include up to 75000 individuals. Approximately 64 % are adult bees, 21 % are pupae, 10 % are larvae and 5 % are eggs (BODENHEIMER, 1937; FUKUDA, 1983).

The underlying, fundamental social structure of a honeybee colony is that of a matriarchal family (SEELEY, 1985). One long-lived female, the queen, is the mother of the members of a typical colony in summer.

Most of her offspring are workers, daughters which never mate but are able to lay unfertilized eggs, which develop into males (drones). In a queenright colony, a considerable proportion of the drone eggs can be laid by workers, however, most of them are usually eaten by other workers (RATNIEKS & VISSCHER, 1989).

Queens control the gender of their offspring by laying unfertilized, haploid, or fertilized, diploid eggs. Males develop from haploid eggs while females develop from diploid eggs (KERR, 1969; MICHENER, 1974; CROZIER, 1977). Nevertheless, it seems that workers can also influence the sex of the offspring by supplying the right size of cells to the queen to lay eggs in (KOENIGER, 1970). Drones are raised in slightly bigger cells than workers (HEPBURN, 1986). In exceptional cases diploid eggs which are homozygous at the sex locus (BEYE ET AL., 2003) can develop into sterile drones, but since they are generally eaten by worker bees in the larval stage, virtually all males are haploid in natural populations (WOYKE, 1963; ROTHENBUHLER ET AL., 1968).

Whether an egg that is diploid and heterozygous at the sex locus develops into a worker or a queen depends on the composition of food given to the developing bee during the first three days of her larval life (HAYDAK, 1970). The difference in the combination of nutrients is the concentration of hexose sugars (BUTLER, 1960; HANSER & REMBOLD, 1960; REMBOLD, 1973). Evidently, the sweetness triggers different larval feeding rates, different levels of juvenile hormone during development, and ultimately different developmental programs for the two types of female bees (BEETSMA, 1979; DE WILDE & BEETSMA, 1982).

The level of reproductive division of labor between the two female castes is so advanced, that literally a honeybee queen after performing her nuptial flights, functions as little more than an egg-laying machine (SEELEY, 1985). The other tasks inside and outside of the hive are performed by the workers.

The drones are unable to fulfill the multiple tasks of a worker. They lack morphological structures like wax glands, hypopharyngeal glands, sting and poison glands as well as pollen-collecting hairs. In addition, drones are literally handicapped even if it comes to most elementary social interaction like feeding hive mates (HOFFMANN, 1966). Their mouthparts are too short for transferring food; therefore they are unable to pass regurgitated food to other bees. The short live of a drone is solely oriented towards its chance of reproduction.

1.3.3 Division of labor, task allocation and life span in the honeybee

Division of labor among the workers is central to the social organization of honeybees and is fundamental as it allows the colony to operate far more efficiently than if it were a simple aggregation of identical individuals (WILSON, 1985A).

A central question is how the activities of individual workers are integrated to enable the continuous development and reproduction of colonies despite changing internal and external conditions. The regulation of age-based division of labor among workers demands a high level of colony integration.

Generally, adult bees perform a variety of tasks in the hive that depend on several factors: the season, the bee's age, its past experiences, the current age demography of the colony and the current demands of the colony (RÖSCH, 1925, 1927, 1930; SCHMICKL & CRAILSHEIM, 2004).

Honeybee workers show a distinct bimodal longevity distribution in temperate zones and may be classified either as short-lived summer bees or long-lived winter bees (MAURIZIO, 1950). Bees emerging in spring have an average lifespan of about 25 to 35 days, whereas winter bees normally live for 6 to 8 months (MAURIZIO, 1950; FREE & SPENCER-BOOTH,

1959). The subchapters 1.3.3.1 and 1.3.3.2 focus in detail on some of the various factors influencing the age and physiology differences between summer and winter bees.

1.3.3.1 Summer bees

During the first 0 to 2 days after emerging from the brood cell, a summer worker bee cleans cells to prepare them for reuse (RÖSCH, 1925; LINDAUER, 1952). The next older age class associated with brood production is the nurse bee. Nurse bees are typically from 5 to 16 days old (HAYDAK, 1963). They digest the pollen and nectar and convert it into a fluid called "jelly" (HANSER & REMBOLD, 1964), secreted by their hypopharyngeal glands which have developed at that age. The jelly is fed to the larvae, the queen, drones and to adult bees performing other tasks (CRAILSHEIM, 1992). Jelly can be mixed with honey in different proportions, that way, worker jelly containing less and royal jelly containing more hexose sugars can be distributed selectively.

From day 11 onwards, the hypopharyngeal glands regress and instead the wax glands become functional for comb building. At the time honeybees become foragers both glands usually have degenerated (RÖSCH, 1930).

Several early studies showed that younger, middle-aged and older worker bees choose among age-characteristic tasks (RÖSCH, 1925; LINDAUER, 1952; SAKAGAMI, 1953). This phenomenon of behavioral change with time is called "age polyethism" (FREE, 1965; SEELEY, 1985).

However, the task schedule depending on age polyethism is not as strict as it is often pictured. SAKAGAMI (1953) states that his repetitions of RÖSCH'S experiments on age-dependent division of labor from 1925 showed few similarities. He states that correlations between performed task and age were only measurable if a "normal" colony was observed in "optimal" foraging conditions. This inconsistency should not be deemed to be a dysfunction in an abnormal colony, or as a measuring fault. In contrast, the observed variation within a colony appears to be the adaptiveness which is essential for responding adequately to the changing environmental conditions and therefore maintaining the fitness of the colony. Indeed, even older foraging workers can return to brood rearing nurse tasks which are normally performed by very young bees, given that the colony conditions require such shifts. SAKAGAMI (1953) showed in his experiments with single-cohort colonies, that even tasks that are strictly related to young bees (like nursing) can be performed by older bees, if the colony only consists of older worker bees, and that certain tasks which are characteristically performed by older bees (like foraging) can be performed by precocious young bees, if the colony only consists of young worker bees. It takes several days until the hypopharyngeal glands of older worker bees have resumed the production of jelly and young bees need some time to adapt to the flying which plays a decisive role for successful foraging. Later studies revealed that the division of labor in honeybees is primarily influenced by the colony's total population (=workforce), by its age distribution and its current ration of brood to nurses (=workload).

WINSTON and PUNNETT (1982) and WINSTON and FERGUSSON (1985, 1986) showed that the total colony population size and not the amount of brood influences the starting age of foraging.

The basic principles for the ability to react flexibly to a changing requirement are physiological regulatory mechanisms which are represented in the activities of glands and the existence of certain hormones (HUANG & ROBINSON, 1996).

1.3.3.2 Winter bees

Winter bees emerge during a restricted period in late summer and autumn, and differ from summer bees with respect to several physiological characteristics. The gland and hormone activity in winter bees is not as linear as it is in summer bees (MAURIZIO, 1950; FREE & SPENCER-BOOTH, 1959; FLURI ET AL., 1982; CRAILSHEIM, 1990; HUANG & ROBINSON, 1995).

If performing many tasks is a characteristic of summer bees, keeping up a few essential tasks with fewer bees is characteristic of winter bees.

Their physiology is mainly oriented towards surviving the cold season. In early fall, when brood is still present, the winter bees perform all tasks summer bees do as well. When there is no more brood to be taken care of and there is no need for foraging, the worker bees in winter are busy keeping a high temperature at the core of the winter cluster (SIMPSON, 1961). Besides that, their main activities are keeping the queen alive by feeding, keeping the core of the winter cluster at moderate temperatures by heating and by distributing the resources the bees collected along the warmer seasons.

When spring arrives colony requirements change and the winter bees have to collect resources and must raise the first generation of summer bees. With changing tasks the physiological requirements of the workers change as well and the activity of the wax and hypopharyngeal glands resumes (CRAILSHEIM, 1990).

1.3.4 Juvenile hormone (JH)

A concomitant factor for division of labor is the hormone level, which is known to regulate the age-dependent task specialization. The highest importance is attached to the juvenile hormone (JH) (FAHRBACH & ROBINSON, 1996; HUANG ET AL., 1991). Juvenile hormones are synthesized and released by the *corpora allata* and play many fundamental roles in the postembryonic physiological and behavioral development of insects (NIJHOUT, 1994). Juvenile hormone III is the only homolog found in worker bees (FLURI ET AL., 1982) and its titer increases as the adult bee ages, from about 5 p/mol per 100µl hemolymph on the first day following eclosion to over 20 p/mol per100µl hemolymph 3 weeks later (FLURI ET AL., 1982; ROBINSON ET AL., 1987) (Fig. 1.4).

However, the increase of JH in the honeybee hemolymph is not steady. JASSIM ET AL. (2000) found that there is a peak of JH titer in 2 to 3 day old adult bees, the significance of which is still unknown (Fig. 1.4). JH titers can also change significantly under stress factors commonly experienced by workers in experimental manipulations (LIN ET AL., 2004). Since the hemolymph titer of JH increases as the honeybee ages, low titers are consequently associated with the performance of tasks in the nest such as brood care during the first weeks, whereas a higher titer at about three weeks of age is associated with foraging.

Treatment with juvenile hormone, juvenile hormone mimic or juvenile hormone analogue is known to induce precocious foraging (JAYCOX ET AL., 1974; JAYCOX, 1976; ROBINSON, 1985, 1992; ROBINSON ET AL., 1987, 1992; SASAGAWA ET AL., 1989). Treatment experiments also indicate that JH is involved in the regulation of age polyethism throughout the bee´s life and not only during the shift to foraging (ROBINSON ET AL., 1987).

1.3.5 Vitellogenin

The protein status appears to be a major determinant of the honeybees' lifespan (MAURIZIO, 1950, 1954; DE GROOT, 1952; SCHATTON-GADELMAYER & ENGELS, 1988; BURGESS ET AL., 1996). Being the most abundant hemolymph protein found in workers and queens, the very high-density lipoprotein vitellogenin seems to play a crucial role for the honeybee (AMDAM ET AL., 2002). It strongly reflects the general protein status of the bee (ENGELS & FAHRENHORST, 1974; CREMONEZ ET AL., 1998). It has been suggested, on the basis of results from workers, that vitellogenin acts as antioxidant to promote longevity in queen bees (CORONA ET AL., 2007).

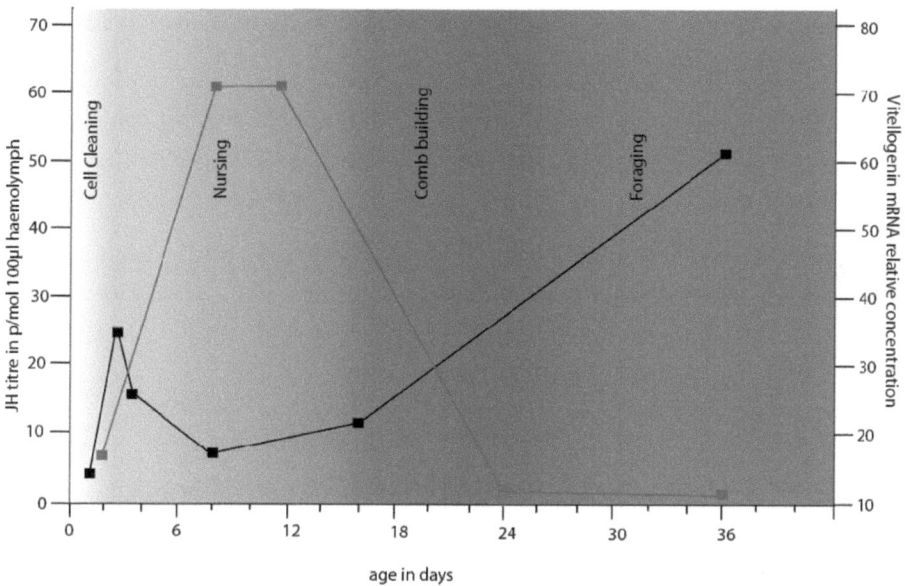

Fig. 1.4 Hormone levels in the honeybee worker in relation with age and labor

The hormone levels of JH (black line) (in p/mol per 100μl hemolymph) after FLURI ET AL., (1982), ROBINSON ET AL., (1987) and JASSIM ET AL. (2000) and Vitellogenin (grey line) (in mRNA relative concentration) after CREMONEZ ET AL. (1998) and ENGELS ET AL. (1990) in the honeybee worker in relation with the age and the labor generally associated. The vitellogenin titer described by ENGELS and FAHRENHOST (1974) shows a similar curve but is measured in the percentage of a fraction in the hemolymph spectra.

Vitellogenin is also a potent zinc (Zn) carrier (FALCHUK, 1998) and the amount of hemolymph zinc is strongly correlated with the vitellogenin level in honeybees (AMDAM ET AL., 2005). Zinc is required as a catalytic, structural and regulatory ion. Zinc deficiencies are known to induce oxidative stress and apoptosis in several cell lines in mammals, including nerve and immune cells (MOCCHEGIANI ET AL., 2000).Vitellogenin is produced by the fat body of many insect species, and is generally described as a female-specific hemolymph storage protein, a yolk glycoprotein that is secreted into the hemolymph and taken up by developing oocytes (HAUNERLAND & SHIRK, 1995). The rate of vitellogenin synthesis is negligible when the worker emerges, but increases rapidly within 2 to 3 days and may be enhanced when the worker starts nursing (ENGELS ET AL., 1990) (Fig. 1.4). The protein status of a worker is mainly given by the amount of protein present in its fat body, hemolymph and hypopharyngeal glands. The fat body builds up during the first days of adult life (KOEHLER, 1921; HAYDAK, 1957).

In summer, the maximum amount of proteins in the fat body of a worker bee is obtained after approximately 12 days (Fig. 1.4), while it may increase far beyond this level over an extended time period in late autumn (MAURIZIO, 1954; FLURI & BOGDANOV, 1987).

The amount of proteins decrease during winter, and spring levels may be even lower than the quantities found in summer foragers (MAURIZIO, 1954; FLURI & BOGDANOV, 1987). Wintering workers have, in general, a high hemolymph vitellogenin titer, but it is higher in late autumn than at the end of winter (FLURI ET AL., 1982).

1.3.6 Genetic influence on the division of labor

Genetic predisposition is another influence on the division of labor in a colony (ROTHENBUHLER & PAGE, 1989; PANKIW & PAGE, 2001). The genetic structure of honeybee colonies is complex, because queens mate usually with about 5 to 30 different drones (ADAMS ET AL., 1977; ESTOUP ET AL., 1994; FUCHS & MORITZ, 1999; NEUMANN ET AL., 1999; NEUMANN & MORITZ, 2000) and in egg fertilization use the sperm of at least several drones at any one time (PAGE, 1986). Therefore, each colony consists of numerous subfamilies, each of which is a group of super-sisters (r=0.75) (PAGE & LAIDLAW, 1988).

Behavioral differences among members of different subfamilies were demonstrated for guarding and corpse removal (ROTHENBUHLER, 1958; ROBINSON & PAGE, 1988) and for foraging and nest-site scouting in honeybee colonies (ROBINSON & PAGE, 1989).

Within queenless colonies, subfamily differences have been found for the exchange of food, oviposition behavior, oophagy and drone larval care (MORITZ & HILLESHEIM, 1985; HILLESHEIM ET AL., 1989; ROBINSON ET AL., 1990). The preference of super-sisters under queenless conditions is not supportive for the colony as such, but supports only its own gene pool.

In addition, HELLMICH and ROTHENBUHLER (1986) described different genetic lines, one that regulates pollen stores at high level and another that regulates them at low level. But both lines exhibited demand-driven regulation when brood periods were compared with broodless periods. The rate of usage of pollen was the same for both lines. Like age-dependent division of labor, genetic predisposition shows correlations in ecologically well-balanced times, but when an increased workforce is required for some duties.

1.3.7 The nest of the honeybee

Honeybee colonies under natural living conditions inhabit hollow trees. Nest cavities are vertically elongated and approximately cylindrical, having approximately the shape of the cavity they are built in (SEELEY & MORSE, 1976). A honeybee nest consists of a set of combs organized in a characteristic manner facilitating proper thermoregulation of the brood area (HIMMER, 1932; VILLA ET AL., 1987; SOUTHWICK & HELDMAIER, 1987) and featuring a storage place for the precious honey.

The middle of the comb usually contains an area filled with brood, surrounded by empty cells, so the brood nest can grow. The brood nest in turn is enclosed in a ring of cells containing pollen, the easiest transport to bring the pollen into the brood nest, allowing the nurses to gain access to the pollen supply without leaving the brood (CAMAZINE, 1991).

The remaining upper part of the comb is filled with honey or nectar.

In contrast to early assumptions that this special organization is somehow dictated by the queen, CAMAZINE (1991) described simple individual processes that can result in the observed spatial organization. SEELEY (1982) shows the strong interdependence of task performance and the spatial distribution of task-associated workloads: workers of one age-class perform a variety of tasks, which are mostly localized within the same region in the hive.

This functional separation of the nest into honey bearing regions and brood and pollen bearing regions is associated with several differences in comb structure:

The brood comb is dark brown or even black, because the pupal skin remains at least partly in the cell after the larval moult and darkens the brood cells by degrees. After a while, the skin remains change the consistency of the brood comb to an almost parchment-like texture. The cell pattern is regular, meaning the cells are arranged in straight, horizontal rows, the cell walls are straight and the cross section between the cells is regularly hexagonal (PIRK ET AL., 2004). The comb width is uniform, either 21 to 24mm (worker brood cells) or 25 to 29mm (drone brood cells) wide, but varies between subspecies (HEPBURN, 1986).

The honeycomb by contrast is light yellow to light brown. The comb width is irregular; the cell sizes are of various diameters and depths. The cell walls are often curved, the cell pattern is often irregular and the cross section is often irregularly hexagonal (HEPBURN, 1986). Therefore, brood comb and honeycomb are not only different in their cell contents but also in shape, texture and color of the cells.

SCHMICKL and CRAILSHEIM (2004) assume that the brood nest itself plays an important role in the ability of honeybees to regulate proper nest homeostasis. It is the center of the trophallactic network. Even though they believe that nest homeostasis is decentralized and self-organised, they argue that the major part of these self-organisational processes operate within a distinct area of the brood nest.

Task location efficiency might be of great importance. If the tasks performed concurrently also co-occur spatially in the nest, then the mean free path between tasks should be minimized, and this should help maximise efficiency in locating tasks. To test this spatial-efficiency hypothesis one can investigate whether the task-set for each age of pre-foraging workers maps onto a specific nest region, or in contradiction to the hypothesis, onto spatially segregated sites about the nest (SEELEY, 1985). The centralization of the brood nest and its need for thermoregulation is a text book example for the spatial-efficiency hypothesis. If there were two separate, smaller brood nests or single brood cells distributed over the comb, the energy expense would rise.

1.3.8 Thermoregulation and heating activity

Surviving the winter as a colony and raising brood in spring is only possible because honeybees have developed mechanisms to keep the hive and the brood at temperatures that are necessary for the insects to survive the winter and for their pupae to develop into fully functional worker bees. The ability of honeybee workers to generate large amounts of heat through so called "shivering thermogenesis" (STABENTHEINER ET AL., 2003) depends to a large extent on the glycogen metabolism (PANZENBÖCK & CRAILSHEIM, 1997).

Because of this attribute, HEINRICH (1993) describes the honeybee as a highly atypical flying insect. *"They seek the warmth of their companions in the nest and are unavoidably subjected to heating them"*.

Introduction

Some aspects of the honeybees' behavior and physiology are probably also shaped by the constant access to food. They normally have energy supplies constantly within reach, and so they do relatively little to conserve them (HEINRICH, 1993). Honeybees generally stay endothermic as long as they have sugar in their honey stomach or midgut. When the food is gone, they soon exhaust their tissue reserves and die. What seems like a handicap for the bee is usually no problem, because a honeybee worker is only solitary and without direct access to food while foraging. In the hive there is ether enough honey or nestmates ready to feed each other within a very short distance. Since honeybees use mostly sugar as an energy substrate for muscular activity (JONGBLOED & WIERSMA, 1934; LOH & HERAN, 1970; SACKTOR, 1970; ROTHE & NACHTIGALL, 1989), the level of glycogen in the hemolymph must be kept high to provide an adequate fuel supply for the heat-generating flight muscles (CRAILSHEIM, 1988) which are the most metabolically active tissues known (SOUTHWICK & HELDMAIER, 1987).

In honeybees, food is stored in the crop or "honey stomach". Such a crop occurs in other hymenoptera species as well. HÖLLDOBLER and WILSON (1990) frame it "the social stomach", because its contents are only used to a certain part by the individual, since the food can be regurgitated and fed to other individuals of the colony.

The crop contents of a honeybee never enter the bloodstream directly. The crop wall is impermeable for water and sugar. Liquids stored in the crop have to pass a sphincter muscle, the ventriculus, which works as a valve that can release food into the midgut, where it is transferred into the bloodstream (BLATT & ROCES, 2001).

A crop load of sugar solution can provide a bee with food for several hours. But even inactive, caged honeybees with a full crop held at room temperature die within 7 hours after being separated from their food source (HEINRICH, 1993). A physiologically challenging activity like flying or heating will consume their sugar fuel even faster, so the crop content and its sugar concentration consequently reflects the demand of the upcoming task (NIXON & RIBBANDS, 1952; CRAILSHEIM, 1988). SOTAVALTA (1954) reported that honeybees he kept flying for 10 to 15min died 5 to 10min later unless food was given to them.

Honeybees must generate heat not only for brood heating and in the winter cluster, but to warm up prior to flight if air temperatures are low. As in all other insects, heat is generated by the flight muscles during shivering. Wing and thoracic vibrations are generally not visible to the naked eye and the heating bees may appear to be quiet and "at rest".

Introduction

The thoracic heat is a by-product of flight during which up to 60 % of the energy is released as heat or as JOSEPHSON (1981) put it: *"It [Insect flight] efficiently converts chemical energy to mechanical power and, because of biochemical inefficiencies, heat."*

The elevated energy consumption in brood heating and flying can be concluded from the equality in oxygen consumption by the bees for both activities, which is 1.16µl/g/min during flight muscle shivering and 1.14µl/g/min during flight (HEINRICH, 1993).

The relatively small mass of honeybees means that the passive-connective heat loss is very rapid for a solitary individual. But having evolved a highly social system with tens of thousands of individuals sharing a nest, they have reduced heat loss as a group by building a cluster whenever necessary. Individual bees have only a limited capacity to stabilize their thoracic temperature and individual thoracic temperatures generally fluctuate (HIMMER, 1925, 1927; ESCH, 1960; HEINRICH, 1981A), but when bees are gathered together in larger groups (FREE & SPENCER-BOOTH, 1958) body temperature stabilization becomes ever more precise because of the reduced thermal inertia of the larger mass. SOUTHWICK and HELDMAIER (1987) wrote that the efficiency of tight clustering in winter can reduce the effective area of heat exchange by as much as 88 %.

Thermal performance of honeybees is correlated not only with season but also with age. Young bees only gradually develop the capacity for endothermic heat production (HIMMER, 1932; ALLEN, 1955; HARRISON, 1986; STABENTHEINER & SCHMARANZER, 1987). Before they have developed the capacity to generate heat by shivering, new workers tend to stay in the warm brood nest (FREE, 1961). Within the first few days the maximal thorax-specific metabolic rate closely corresponds to the increase in enzyme activities. Pyruvate kinase and citrate synthetase activities increase (tenfold) up to only 4 days of age, and then gradually decline (HARRISON, 1986).

By contrast, BUJOK (2005) demonstrated that worker bees show proper brood heating activity 48h after eclosion even though their physiology should not be fully adapted to this task. Since young bees kept in cages outside the hive and without direct access to a queen show signs of higher JH activity which is known to have a potent effect on muscle growth, the flight capability (WYATT & DAVEY, 1996) and the respiratory metabolism (NOVAK, 1966) in insects, both findings are not mutually exclusive.

The nest of social bees serves as incubator for raising offspring and as refuge from enemies and temperature extremes. The importance of this rigidly controlled microclimate cannot be overemphasized in any study of social insects. Indeed, most treatises on the

Introduction

social life of insects discuss numerous facets of this fascinating thermoregulatory behavior at length.

Like other Apis species, the European honeybee with its nearly world-wide distribution, probably originated in the tropics. The perennial nature of its colonies and its reproduction by swarming are common features among tropical social bees and distinguishes it from most social bees endemic to cold temperate regions. Winter is still the time of greatest mortality of even those races now adapted to northern climates, and colony thermoregulation is a critical feature of its biology. Indeed, beekeepers usually report of about 10 % colony mortality of all colonies in Middle Europe every winter (OLDROYD, 2007).

It is not surprising that no area of insect thermoregulation has received as much attention as honeybee nest-temperature regulation, since it was discovered over two centuries ago by RÉAMUR (1742) and HUNTER (1792). HUNTER originally suggested that the warmth generated by the bees kept the wax soft so as to allow them to shape it into cells. The ductility of beeswax is indeed uniquely optimized at the temperature that is regulated within the nest (HEPBURN ET AL., 1983).

As any homoeothermic organism, the metabolic rate of bee groups increases at decreasing air temperatures (WOODWORTH, 1936; ROTH, 1965; HEINRICH, 1981A,B; SOUTHWICK, 1982, 1983, 1985; SOUTHWICK & HELDMAIER, 1987).

In large swarms most bees in the deep interior of the cluster are shielded from low temperatures. These bees are unavoidably warmed by the dense crowding, and they cool slowly. Indeed, based on cooling rates of bees inside the cores of heated dead swarms, calculations of how much heat core bees need to produce if they were shivering to keep warm indicates that even their resting metabolism is about ten times more than needed to keep warm at air temperatures near 0 °C (HEINRICH, 1981A,B). In other words, live swarms are unavoidably heated and have active mechanisms of dissipating heat from the core.

One of the major responses of bees that are unrelated to their individual behavior relates to brood. A dramatic change occurs in the colony response after brood rearing begins in spring. Brood rearing can occur at air temperatures from -40 °C to 40 °C or more. Over this wide range of temperature the bees maintain the temperatures of the brood nest between 33 °C and 36 °C by heating or cooling (HIMMER, 1927; SEELEY & HEINRICH, 1981; ESCH & GOLLER, 1991; HEINRICH, 1993). If temperatures are not kept within these limits, the results may be shrivelled wings and other malformations (HIMMER, 1927), as well as brain damage and losses in behavioral capability (TAUTZ ET AL., 2003; GROH ET AL., 2004).

Introduction

In the absence of brood such as in swarms or in the hive in fall and winter, the temperature of the bees at the nest periphery must not fall below the chill-coma temperature of near 10 °C (FREE & SPENCER-BOOTH, 1960).

During brood incubation, the bees shiver where they might otherwise allow their thoracic temperature to decline (RITTER, 1978; KRONENBERG & HELLER, 1982). The details of thermoregulation of brood care are not clear, but it is certain that bees are attracted to clusters of capped brood (KOENIGER, 1978; RITTER & KOENIGER, 1977), where they have a higher metabolic rate than at combs with honey (KRONENBERG, 1979; KRONENBERG & HELLER, 1982).

The heating bees station themselves on the brood comb where they transfer heat either by pressing their hot thoraces onto capped cells (BUJOK ET AL., 2002), or by crawling head first into empty cells within the brood comb to heat neighbouring brood from the side (KLEINHENZ ET AL., 2003). This uninterrupted cell-heating activity was observed to last up to 32.9 min by KLEINHENZ ET AL. (2003).

The cues that cause both attraction and shivering may be both chemical and tactile (HEINRICH, 1993). It is not known, if the bees respond directly to brood temperature. If there is a tight coupling between their own thoracic temperature and that of the brood, then they could potentially regulate their own thoracic temperature in the presence of brood so that brood temperature regulation results secondarily.

BUJOK (2005) showed that honeybees which are confronted with frozen dead and reheated brood (to 35 °C) show normal heating behavior for several days. The evidence is that the presence of brood and heat is necessary to trigger brood heating activity. There is more evidence against the hypothesis, that brood heating is a by product of regulating the own body temperature. BUJOK (2002) found that heating bees not only station themselves on the capped brood but they touch the caps of the brood cells with the tips of their antennae. Honeybee workers have temperature receptors on the last five antennal segments whose impulse frequently increases with decreasing temperatures (VON LACHER, 1964). This coherence suggests that a higher thoracic temperature, pressing the thorax on the capped brood and "checking the temperature of the cell cap" are not coincidences.

Besides flight, brood heating and the winter cluster higher thoracic temperature is a sign of aggression in honeybees as well. Elevated body temperature is a signal for aggressive behavior prior to fight or flight in many animal species. Especially in insects, where thoracic muscles need a certain "operating temperature", body heat is necessary to react to any kinds of stress threatening the individual or the colony. The attacking temperature in honeybees is higher than the temperatures measured in clustering or flying (ESCH, 1960; HEINRICH, 1971; ONO ET AL., 1987, 1995; STABENTHEINER, 1996; KASTBERGER & STACHL, 2003; KEN ET AL., 2005).

The interrelationship of aggressive behavior and thermoregulation in *A. m. carnica* was described precisely by STABENTHEINER ET AL. (2007). They found that guard bees, foragers, drones and queens were always endothermic, i.e. had their flight muscles activated, when involved in aggressive interactions. Guards make differential use of their endothermic capacity. Mean thoracic temperature was 34.2 °C to 35.1 °C during examination of worker bees but higher during fights with wasps (37 °C) or attack of humans (38.6 °C) They cool down when examining bees whereas examinees often heat up during prolonged interceptions (up to 47 °C) (STABENHEINER ET AL., 2002, 2007). It is hypothized that they do this to enhance chemical signalling via an increase in vapour pressure of chemicals from their surface involved in nestmate recognition (STABENHEINER ET AL., 2007).

The usually not aggressive honeybee queen is endothermic in fights with other young queens and the attack of their cells before they emerge (STABENHEINER ET AL., 2007).

Wasps are particularly dangerous enemies of honeybees and guard bees often attack them directly at the nest entrance. When guard bees are not able to defend such intruders on their own, they recruit other bees to help them (ONO ET AL., 1987). During such mass attacks worker bees of the species *Apis cerana* and *A. dorsata*, increase their thoracic temperature to 45 °C to 48 °C in an attempt to kill the engulfed insects by heat. The heat tolerance of *A. cerana* and *A. dorsata* allows them to survive for temperatures up to 50.7 °C, while the wasps die at 45.7 °C (ESCH, 1960; ONO ET AL., 1987, 1995; STABENTHEINER, 1996; KASTBERGER & STACHL, 2003; KEN ET AL., 2005). This so called thermal killing or "heat-balling" is more or less pronounced in all *Apis* species (KASTBERGER & STACHL, 2003). Even though the European honeybee *A. mellifera* does not engage in excessive "heat-balling", it survives temperatures up to 51.8 °C. One subspecies of the European honeybee, the Cyprian honeybee *A. m. cypria*, is known to suffocate hornets of the species *Vespa orientalis* in a tight cluster (PAPACHRISTOFOROU ET AL., 2005). The honeybee ball around the wasp heats up but only reaches a core temperature of 44 °C. The upper lethal temperature of the hornet is 50 °C. The Cyprian honeybee ball does not "fry" but suffocate the hornet. The honeybees literally squeeze the hornet's breath away by blocking the movements of the tergites (PAPACHRISTOFOROU ET AL., 2007).

1.3.9 Trophallaxis

The food intake to fuel the activity of a honeybee is either done by the individual itself, i.e. it is taking up food while foraging or from the storage. Another possibility is to get fed by another individual which regurgitates food from its crop and transfers it mouth to mouth. This feeding activity between two individuals is called trophallaxis (WHEELER, 1928; FREE, 1956) (Fig. 1.5). This mouth to mouth transfer of food occurs frequently among workers of honeybee colonies.

They share the contents of their crops and sometimes the product of their head glands. Trophallactic interactions can be seen non-randomly between all members of the colony. Their occurrence and success depend on factors such as sex and age of the consumers and donors. Availability and quality of food, time of day weather and season are known to influence this behavior as well (CRAILSHEIM, 1998).

There are two ways a trophallactic contact in honeybees can start: Firstly, a bee can beg for food by extending its proboscis and thrusting its tip towards the mouthparts of another bee, termed the donor if the contact leads to a transfer. If the begging bee is successful, it is termed a recipient, while the donor bee responds by regurgitating food and thereby is initiating a trophallactic contact (Fig 1.5). Secondly, a bee can offer food without being stimulated directly by another worker, by opening its mandibles and moving its still-folded proboscis slightly downwards and forwards from its position of rest. A drop of regurgitated liquid food can often be seen between the mandibles and on the proximal part of the proboscis (FREE, 1959). If a recipient bee touches that droplet with its antennae and then thrusts its proboscis between the mouthparts of the donor, this also results in a trophallactic contact (MONTAGNER & PAIN, 1971) (Fig 1.5).

While engaging in a trophallactic contact, the antennae of both individuals touch each other frequently (ISTOMINA – TSVETKOVA, 1960).

If the antennal contact is hindered by partial or total amputation, the success of transfer is reduced. Pioneering experiments concerning the role of antennae in trophallactic activities of honeybees were performed by FREE (1956). He amputated different parts of one or both antennae. His experiments showed that the abscission of the antenna reduces all feeding activities: the more segments were affected, the less feeding activity was measurable. Especially the initiation of the trophallactic contact seems to be connected to the antennae. The underlying antennal motor pattern of the recipients is not inherent. An adult honeybee, only a few hours old, does not extend the antennae towards the donor, but extrudes the proboscis. The antennae are used by and by until after five or six days the young bee progressively acquires the antennary ritual of solicitation: the frequent and reiterated introduction of the extremity of one or of the two antennae between the mandibles of the begged bee (MONTAGNER & PAIN, 1971). These findings correspond with the observations of FREE (1959) who showed that behavior patterns associated with food transfer are innate, but lack the precision and co-ordination of older workers in newly emerged worker bees or in individuals that have been kept in isolation for several days.

Fig 1.5 Trophallactic contact in *A. mellifera* (Drawing by R. Basile)

The donor (left) opens its mandibles and regurgitates a droplet of fluid which is supported by the proboscis, while the recipient (right) thrusts its proboscis between the donors spread mandibles.

The importance of the antennae in releasing food transference and in helping bees to orientate their mouthparts to one another is probably the reason, why bees have difficulty in feeding each other through a wire-gauze screen whose mesh is below a certain size, even though they are able to insert their tongues through it (FREE & BUTLER, 1958).

Not all trophallactic contacts in the hive are feeding contacts. The transferred fluid can consist of honey, nectar, water or jelly in different quantities. Especially protein-rich food is passed from nurses to larvae or to workers and drones in need of jelly.

Newly hatched drones are fed extensively with jelly by nurse bees. Drones solicit food from workers and from other drones (OHTANI, 1974), but trophallaxis between drones has never been observed because they lack the ability to pass the regurgitated food from their mouthparts (HOFFMANN, 1966).

All of the queen's nutritional requirements are given to her by bees in her court via trophallaxis (ALLEN, 1960; FREE ET AL., 1992). A queen can also survive isolated and feed herself (WEISS, 1967) but this situation is only reported if a queen is not yet mated (BUTLER, 1954) or not laying eggs at the moment (PREPELOVA, 1928). Although a queen normally receives food only, BUTLER (1957) found that when he introduced queens into strange colonies i.e. a colony she was not raised in, they sometimes adopted a submissive attitude and offered food to worker bees of the recipient colony in a similar manner to that is shown by submissive workers (BUTLER & FREE, 1952; SAKAGAMI, 1954; MEYERHOFF, 1955)

A transfer from one worker to another can last from less than one second up to some min (ISTOMINA-TSVETKOVA, 1960; KORST & VELTHUIS, 1982). A transfer can be very fast. Maximum speed of transfer was observed by FARINA and NUNEZ (1991) with 1.6µl per second. But as observed in cage experiments by KORST and VELTHUIS (1982), even if the attempts last more than 10 seconds, they are not necessarily successful.

Therefore, the duration of a trophallactic contact is not inevitably related to the transferred amount of liquid. Brief feeding contacts often lead to discussion as to whether they can be counted as real trophallactic transmission at all. FARINA and WAINSELBOIM (2001A) observed trophallactic contacts with a thermal imaging camera. Since the body temperature of a honeybee is highly variable (between the present ambient temperature and 45 °C), the food transmission creates a contrast in the thermal picture, if it is transferred from one individual to another with a different temperature (WAINSELBOIM & FARINA 2001A). The warmer or cooler fluid of the transferred food "tints" the proboscis of the recipient. So not only the presence of liquid food but the direction of the flow can be determined easily.

During periods when the colony needs a lot of water, some foragers specialise in water collection (Seeley & Morse, 1976; ROBINSON ET AL., 1984; KÜHNHOLZ & SEELEY, 1997). This water is transferred to other bees in the hive by trophallactic contacts as well (PARK, 1923; VON FRISCH, 1965).

The trophallactic interactions – the donation and the reception of food from one bee to another – is an important factor in making the complex social community work (FREE, 1959) and is often attributed of being the origin of sociality itself (SLEIGH, 2002).

Introduction

The usually non-aggressive feeding behavior in honeybees is unequal compared to what happens, for example in wasp society, in which each worker begs in an individual contact for regurgitated food for itself and shows aggressive behavior towards the individual that refuses to regurgitate (MONTAGNER & PAIN, 1971). This important difference may account for the annual character of the wasp society and the perennial nature of the bee society.

Conflicts in the context of trophallaxis as they are described for social wasps and other social hymenopterans can occur between honeybee workers as well. If for example foreign bees enter a colony or a bee gets under attack in a cage experiment, the defeated or subdominant individual often regurgitates food. Food offers as appeasing gestures are well known in social and non-social insects and even in vertebrates (Social *Vespidae* - HUNT, 1991; Porine ants - LIEBIG ET AL., 1997; Hallictine bees - KUKUK & CROZIER, 1990; Carpenter bees - VELTHUIS & GERLING, 1983; SWEAT bees – WCISLO & GONZALES, 2006; Bonobos - BLOUNT, 1990)

In honeybees the food regurgitation of submissive individuals is generally considered as appeasing gesture (BUTLER & FREE, 1952; SAKAGAMI, 1954; MEYERHOFF, 1955; BREED ET AL., 1985), because there is a correlation between individual worker dominance and trophallactic behavior (HILLESHEIM ET AL., 1989). The advance a dominant individual gains by receiving food from subdominant workers makes it more probable that these dominant bees can develop ovaries and become reproductive egg layers. Such positive correlation between trophallactic dominance and developing ovaries in honeybees is shown by KORST and VELTHUIS (1982), LIN ET AL. (1999) and by HOOVER (2006)

The trophallactic activity in the hive is influenced by many factors. The location of the bees in the hive is of particular importance. The more bees share one area, the higher are the chances of meeting and engaging in a trophallactic act. The brood comb is usually the area of highest honeybee density in the hive and most of the trophallactic contacts are therefore observed in this area are (SEELEY, 1982).

Bees of similar age seem to feed each other preferably, this might derive from the fact that bees of the same age often perform similar tasks and therefore might be an accompaniment of spatial distribution and age polyethism. FREE (1957) was able to demonstrate that bees of all ages feed partners of all ages, but there is a preference to feed bees of a similar age. The only exceptions were freshly emerged bees and one day old bees that did not donate food to any considerable extent, but received it as frequently as older age hive mates.

Introduction

Contradictory results were published by PERSHAD (1966), who showed that 2 to 4 day old bees are potent donors and MORITZ and HALMEN (1986), who found one day old bees and bees between 15 and 20 days to be the most active donors.

These differences might be caused by the different way the experiments were conducted. The caging, the different amounts of food and the various group sizes could have influenced the results a lot, because the trophallactic behavior in honeybees is not only influenced by the individual honeybee itself, but also from factors like sugar concentration, flow rate at the feeder and previous occurrences concerning food flow or quality.

FARINA and NUNEZ (1995) showed a dependency of donating contacts on the volume in their crops and on the concentration of previously ingested sucrose solutions.

Although in general worker bees about to donate food have a fuller crop than those about to receive it, there is a considerable overlap in the amount of food in the crop of bees of these two categories (FREE, 1957). Whether a bee offers, or begs for food may be influenced by many factors and is not governed entirely by the amount of food in its crop. Attempts have been made to analyse the stimuli to which a worker responds when it offers or begs for food (FREE, 1956). It was found that both types of behavior are directed more to the head than to any other part of a bee's body and that even an excised head is sufficient to elicit both behavioral reactions.

The odour of a head is a most important stimulus, and bees responded more to heads belonging to their own colony (VON FRISCH & RÖSCH, 1926; NIXON & RIBBANDS, 1952) than to heads of bees belonging to another colony. Bees sometimes even begged from "model" heads which consisted of small balls of cotton wool which had been rubbed against bees´ heads and had presumably acquired something of their odour (FREE, 1959).

Temperature or changes in temperature seem to be an important factor in trophallactic contacts as well. PERSHAD (1967) measured trophallactic activity of honeybees at temperatures of 23 °C, 31 °C and 37 °C. Feeding activity was highest at 31 °C and lowest at 37 °C within the first 24h of incubation.

ARNOLD ET AL. (1996) examined the cuticular hydrocarbon composition in subfamilies of workers and demonstrated sufficient variability and genetic determinism to suggest they could be used as labels for subfamily recognition. MORITZ and HILLESHEIM (1990) proved the ability of donors to discriminate and to prefer related over unrelated bees. In nature, such situations occur, when bees have drifted from one hive to another.

Introduction

Trophallactic contacts measured in a drifting experiment from PFEIFFER and CRAILSHEIM (1997) showed no difference between contacts of bees that had drifted and bees that had not drifted. MORITZ and HEISLER (1992) demonstrated the ability to discriminate even between half and super sisters in a trophallactic bioassay. Nevertheless, the importance of the ability to discriminate and possibly prefer closely related bees over less or even unrelated bees in natural selection in not yet clear (OLDROYD ET AL., 1994).

The ecology of the honeybee society undergoes stages in which the worker bees are less related to the upcoming generation (if a new queen is raised for example). In such a case, discrimination of workers against less related sisters could affect the nutritional and informational flow of the hive negatively, or even account for its collapse.

The importance of trophallaxis is still not understood in every detail. It is unclear how much the transfer of enzymes via trophallaxis contributes to the ability of freshly emerged bees to digest honey. The drastically reduced level of amino acids in the hemolymph of workers that were kept in an incubator after eclosion indicates that there are substances they need to develop properly, but which they cannot find on the comb (CRAILSHEIM & LEONHARD, 1997, CRAILSHEIM, 1998). Most likely these substances are transferred by older workers to the young bees via trophallaxis, giving these feeding contacts another nutritional and more physiological value and revealing a new task in the hive.

1.3.10 The morphology of the antenna

The antennae of the honeybee workers are of utmost importance for trophallactic contacts. ISTOMINA-TSVETKOVA (1960) showed that honeybees touch each other frequently while exchanging food and FREE (1956) even found a correlation between the number of trophallactic contacts and the number of segments left on the antennae.

A worker's antenna consists of three parts: a basal scape, a pedicel and flagellum comprising ten annuli or segments. The surface of the antenna is packed with sensory receptors, so called sensilla in enormous numbers. The sensilla of the antennae of the worker bee can be classified into ten morphologically distinct types and are sensitive for different types of stimuli (ESSLEN & KAISSLING, 1976) (Tab. 1.1).

Sensilla are spread all over the antennae in different quantities. Most sensilla accumulate at the distal part of the antenna. The tip of the antenna ends in a backwards and outwards looking oval area, the so-called "Sinnesplatte" (MARTIN & LINDAUER, 1966).

Introduction

Apart from their sensory function, the antennae play a significant role in a honeybee's social life and the fulfilment of its tasks in the hive.

Reception of stimuli	Sensillum
Tactile (mechanoreception)	S. trichodeum B1
	S. trichodeum B2
	S. trichodeum D
Chemical (chemoreception)	S. basiconicum
	S. trichodeum D
Olfactoric	S. placodeum
	S. trichoconicum A
Humidity and Temperature	S. coelocapitulum
	S. ampullacerum
	S. coeloconicum
CO_2	S. coeloconicum

Tab. 1.1 Sensilla present on the honeybees´ antennae and stimuli they are receptive to

Experiments, where parts of the antenna were removed, showed that the sensitivity of the antenna and the behavioral performance of the individual are positively correlated to the quantity of sensilla on the respective remaining antennal segments. Not only does feeding activity decline if parts of the antenna are removed, but the brood heating activity declines with a loss of segments as well. BUJOK (2002) showed that the antennae of heating bees are kept in touch with the caps on the brood cells and demonstrated in an additional experiment that losing parts of the antenna correlates positively with a loss in brood temperature.

The chemosensitive or gustatory properties of the antenna are responsible for a response to sugar containing liquids. If the tip of the honeybee´s antenna or parts of the forelegs (KORST & VELTHUIS, 1982) are touched with a droplet of sugar containing fluid (sugar water, honey etc.), the honeybee will extend its proboscis in order to ingest the fluid. The so called PER (proboscis extension response or reflex) is often used in learning experiments (classical conditioning) (KUWABARA, 1957; BITTERMANN ET AL., 1983).

The PER acts as an unconditioned response to the sugar and can be combined easily with a neutral stimulus (smell, patters etc.) becoming a conditioned stimulus, if the PER is carried out without the sugar water after a while. Smell is most often used in PER conditioning, presumably because honeybees are highly sensitive to olfactory stimuli compared to other insects. ROBERTSON and WANNER (2006) found 170 odorant receptors in the honeybee while fruit flies (*Drosophila melanogaster*) only have 62 and mosquitoes (*Anopheles gambiae*) only 79. Their sensitivity to chemicals (chemoreception) is rather low.

In addition, ROBERTSON and WANNER (2006) found only 10 gustatory receptors in *A. mellifera,* compared with 68 in *D. melanogaster* and 76 in *A. gambiae.*

Why honeybees are easy to train with smell to a PER might be because a similar situation can occur in the hive, when a returning forager smells like the resource the recruit is supposed to find. The forager presents a droplet of nectar and smells like the source. After a while the recruit will react to the typical smell of the source with an extension of the proboscis, ready to take over the nectar. The biological purpose of the PER (proboscis extension after sugar has touched the antenna) in the honeybee is not clear yet.

Lately LETZKUS ET AL. (2006) described laterality in honeybee learning performance. Bees were trained to react with a PER (proboscis extension response) to certain olfactory stimuli. The left and right antenna were covered alternately with a silicone compound in order to test their ability to fulfill the learning task by using only one antenna to react to the olfactory stimulus. Bees which had their left antenna covered learned better than bees that had their right antenna covered.

1.4 Specific aim

The central questions of this thesis are how the heating task and the distribution of resources via trophallaxis are managed in the honeybee colony. The main focuses of the different chapters are the initiation of the trophallactic feeding contact, particularly the activity of the antennae in releasing a feeding contact (**chapter two**); the impact of sugar and water content of food to the heating performance of the individual worker bee (**chapter three**); the behavioral differences between donors and recipients of a trophallactic contact on the brood comb (**chapter four**); the regulation of the trophallactic activity on the brood comb and the possible evolution of the performance related reward system which triggers the feeding and heating activity (**chapter five**).

The present work is composed of six chapters. Chapter two to five present different experiments and observation which are written for separate publications in the appropriate international journals.

Chapter two deals with the use of the antennae in releasing a trophallactic contact. The sequence of behaviors performed by the receiver bees at the beginning of a feeding contact includes the contact of one antenna with the mouthparts of a donor bee where the regurgitated food is located. It is known that the antennae are of utmost importance for food exchange in honeybees (FREE, 1956; VON FRISCH, 1965; MONTAGNER & PAIN, 1971; CRAILSHEIM 1998). Even if only parts of the antennae are removed the number of trophallactic contacts decrease strongly (FREE, 1956). Trophallactic contacts were analyzed in respect to the usage of the antennae at the initiation of the feeding activity. Since honeybees show lateralization in learning performance (PER) which involves the antennae (LETZKUS ET AL., 2006), and the antennal movement of the soliciting bees involves only one antenna at a time, the question arises if such asymmetry actually resembles natural conditions as well, e.g. the trophallactic interactions between nest mates, and if this asymmetry recurs in the gustatory responsivity of the antennae.

Chapter three addresses the question how the quality (sugar content) of the ingested food influences the heat production in the honeybee. In several species of the genus *Apis* it has been demonstrated that sugar concentration and foraging distance are factors which modulate the thermal behavior of honeybees (DYER & SEELEY, 1987; STABENTHEINER & SCHMARANZER, 1988; SCHMARANZER & STABENTHEINER, 1988; WADDINGTON, 1990; STABENTHEINER & HAGMÜLLER, 1991; UNDERWOOD, 1991; STABENTHEINER ET AL., 1995; STABENTHEINER, 1996).

In addition, NIEH ET AL. (2006) showed a positive correlation between thoracic temperature and sugar concentration in *Bombus wilmattae* and NIEH and SANCHEZ (2005) a similar effect in *Mellipona panamica*. The study investigates how the sugar and water content of the diet affects the thoracic temperature of worker honeybees under laboratory condition and if there is a positive correlation between sugar and heat as it is described in other hymenopterans.

Chapter four describes a mechanism to replenish the energy resources of the heating bees. Heating bees fulfill the most energy-consuming task in the hive. The separation of brood nest and food storage creates a spatiotemporal gap between brood and food which must be bridged by the heating bees for a necessary reload of honey. The expense of heat loss for a heating bee, which leaves the brood nest, is irrespective of the distance it needs to bridge between the brood nest and the honeycomb. As the food intake in honeybees is related to the task the worker is about to fulfill (NIXON & RIBBANDS, 1952; CRAILSHEIM, 1988) and the energetic requirements for the different activities in honeybees are unequal, the task partitioning system requires a sort of resource management that assures an ideal distribution of the available stocks to the worker bees that are performing the more strenuous activities.

Chapter five investigates the mechanism behind the food distributing system of chapter four (BASILE ET AL., 2008). Heating bees are fed voluntarily by shuttling donor bees. Heating bees never beg for food, they are offered food by the donor instead. How do the donor bees know which bee to offer food to? If the recipient bees in the hive emit a certain signal and therefore initiate an underlying mechanism responsible for the behavior of the donors, the question arises how this signal is produced and how the donors decide to react properly.

Chapter six correlates the finding in a general discussion.

2. Antennal dexterity in honeybees – about the lopsided use of the antennae in trophallactic contacts

2.1 Abstract

Many social insect societies collectively share the resources they gather by feeding each other. These feeding contacts, known as trophallaxis, are regarded as the fundamental basis for social behavior in honeybees and other social insects for assuring the survival of the individual, information exchange among workers, and the welfare of the group. In honeybees, where most of the trophallactic contacts are formed in the total darkness of the hive, the antennae play a decisive role in the initiation and maintenance of the feeding contact. The antennae are used to initiate the trophallactic contact, since they are sensitive to gustatory stimuli.

The sequence of behaviors performed by the receiver bees at the beginning of a feeding contact includes the contact of one antenna with the mouthparts of a donor bee where the regurgitated food is located. For this motor action we found behavioral asymmetry, which is novel among communicative motor actions in invertebrates: honeybees prefer to use of the right antenna in initiating a trophallactic contact. This asymmetry in the preference of the right antenna is without exception, unlike dexterity or sinistrality in humans. The preference of right over left antenna continues, even after removal of the antennal flagellum.

In addition, we found a gustatory asymmetry in the antennae. The right antenna is significantly more sensitive to stimulation with sugar water of various concentrations than the left one.

This present work shows a case for laterality in basic social interaction under natural as well as laboratory conditions and will be relevant for behavioral and neurobiological studies in honeybees as well as in other invertebrates.

2.2 Introduction

The antennae of insects are of the utmost importance in gathering sensory information and are actively used in communication (VON FRISCH, 1967). In honeybees, one of these antennal functions is the initiation and maintaining of a trophallactic contact, in which food and information is transmitted from a donor bee to a receiver bee (FREE, 1956; VON FRISCH, 1965; MONTAGNER & PAIN, 1971; CRAILSHEIM 1998). A trophallactic contact can involve two or more individuals, whereas one acts as donor that regurgitates the food and one or more individuals are the recipients that receive the food (FREE, 1956; CRAILSHEIM, 1998).

There are two ways of how a trophallactic contact can start: Firstly, a bee can beg for food by extending its proboscis and thrusting its tip towards the mouthparts of another bee. If the begging bee is successful, the other bee responds by regurgitating food and thereby is initializing a trophallactic contact. Secondly, a bee can offer food by opening its mandibles, raising the proximal part of its proboscis and regurgitate a droplet of food that is displayed between the mandibles and the proboscis (FREE, 1956). If a recipient bee touches that droplet with its antennae and then thrusts its proboscis between the mouthparts of the donor, this results as well in a trophallactic contact (MONTAGNER & PAIN, 1971).

Pioneering experiments concerning the role of antennae in trophallactic activities of honeybees were performed by FREE (1956). He amputated different parts of one or both antennae. His experiments showed that the ablation of the antenna reduces all feeding activities: the more segments were affected, the less feeding activity was measurable in donors as well as in food receiving bees.

MONTAGNER and PAIN (1971) filmed feeding contacts of newly emerged and older bees and found that young soliciting bees use their proboscis and one antenna to touch the mouthparts of the donor bee, while older bees, perhaps more experienced, touch the mouthparts of the donor with just one antenna for initiating the feeding contact. They interpreted the differences between both groups as a ritual caused by stimulus response conditioning.

The movement of the soliciting bee´s mouthparts at the beginning of the trophallactic contact is directed by the reaction which occurs while being stimulated with sugar water at the antenna. Upon stimulation, the proboscis is extended. This so-called proboscis extension response (PER) is often used in honeybee conditioning experiments. The gustatory sensilla on both antennae are sensitive to sugar water, therefore the PER can be released via stimulation on either one of them (KUWABARA, 1957; BITTERMANN ET AL., 1983).

Similar to the widespread lateralization of the nervous system in vertebrates (ROGERS ET AL., 2002; ROGERS & ANDREW, 2002; VALLORTIGARA & ROGERS, 2005) findings of lateralization in the honeybee numbers amongst the handful of studies showing that invertebrate species may be lateralized as well. Other examples of lateralization in invertebrate species include a side bias seen in spitting spiders, *Scytodes globula*, to probe potential prey with the front legs on the left side (ADES & RAMIRES, 2002) and for this and many other species of spiders and ants to sustain injury to legs on the left side (HEUTS & BRUNT, 2005). An asymmetrical neural structure in the fruitfly brain is coincident with the ability to form long-term memories (PASCUAL ET AL., 2004), which indicates an advantage of lateralization as also found in the domestic chick (ROGERS ET AL., 2004).

Laterality in honeybee learning performance was first described by LETZKUS ET AL. (2006). Worker bees were trained to react with a PER to certain olfactory stimuli. Either the left or the right antenna was covered with a silicone compound in order to test the bees' ability to fulfill the learning task by using only one antenna to perceive the olfactory stimulus. Workers which had their left antenna covered learned better than bees that had their right antenna covered.

So the question arises whether this kind of asymmetry is actually significant under natural conditions as well, e.g. in the trophallactic interactions between nest mates and whether this asymmetry recurs in the gustatory responsiveness to antennal stimulation. If honeybees prefer one antenna over the other, how will they react to an amputation of one antenna and might this behavioral asymmetry be manifested as well on an individual level as it is in humans?

2.3 Material and Methods

The experiments were conducted at the Beestation of Würzburg University from December 2005 to February 2006 from 10a.m. to 14p.m. with honeybees (*Apis mellifera carnica*) under red light conditions in small arenas (wooden boxes with glass cover). The ambient temperature of the room was 20 °C ±1.

The worker bees were taken from four different brood-free colonies headed by unrelated artificially inseminated queens (ten to twelve drones). Due to the fact that the experiments were conducted in winter, there are no age effects to be considered.

In each experimental run, ten honeybees from the same colony were placed in an arena, offered sugar solution ad libitum and filmed with a digital camera (DCR-SR 190 E Sony) for 3 consecutive hours. The filmed footage consists of 252h (84 x 3h).

For our two treatment groups, we immobilized the honeybees on ice and amputated the first five segments of the honeybees´ flagellum either on the right or on the left side.

In order to prevent pseudo-replicates or influences from the colony, the individually marked bees were replaced for each observation period by bees from another colony. The bees moved freely in the arena and could freely choose the angle for their feeding contacts, so no bee stood in place for a long period of time and successive contacts were always made from a new position.

We counted the frequencies of the antennal activity depending on the angles of the bees´ body axes. Positions that bring the soliciting bee and its right antenna nearer to the left side of the donor ("counter-clockwise" position) are stated as sector left (-20° to -170°). Positions that bring the soliciting bee's left antenna nearer to the right side of the donor ("clockwise" position) are stated as sector right (20° to 170°). Positions where both bees are facing each other in a nearly straight line ("six o'clock position") and both antennal tips have approximately the same distance from the donor's mouthparts are stated as sector zero (0° ± 19°) (Fig 2.1).

In a second set of experiments, we tested 25 individually marked honeybees to see whether they had an individual preference for using one antenna over the other, or whether the bees switch their preference from contact to contact. In each experimental run, five honeybees from the same colony were placed in an arena, offered sugar solution ad libitum and filmed with a digital camera (DCR-SR 190 E Sony) for 3 consecutive hours. The filmed footage consists of 30h (10 x 3h).

The PER Experiments were conducted at the Institute for Ecology at the Technical University of Berlin (TU Berlin)[1]. GRS measurements were made in September 2004. The sucrose responsiveness of the PER to antennal stimulation was quantified by measuring the so-called "gustatory response scores" (GRS) (SCHEINER ET AL., 2004) which are defined as the cumulative number of PERs to 7 consecutive stimulations with different concentrations (0 i.e. water, 0.1, 0.3, 1, 3, 10, and 30 % sucrose). GRS were determined for each antenna separately using an ascending series of sucrose concentrations with an

---

[1] These experiments were conducted by S. S. Haupt. His data is printed with his express permission.

interstimulus interval of 2min. In half of the worker bees, the GRS was determined first for the right antenna, in the other half for the left antenna first. One experimenter stimulated a set of animals while a second experimenter recorded the results. We used half scores (0.5) for a clear proboscis movement that did not result in a fully extended proboscis and full scores (1.0) for a fully extended proboscis upon stimulation. We also calculated a score difference for each individual consisting in the difference of the scores between right and left antenna. A positive score difference corresponds to a higher score on the right antenna.

All statistical analyses were performed using the statistical package Statistica 8©.

2.4 Results

The observation of the antennal use in soliciting honeybees showed a significant preference for using the right over the left antenna for touching the mouthparts of the donor bee and consequently for receiving food from the donor. In 66 % (395 in numbers) of all observed cases, the soliciting bees used the right antenna and in 34 % (204 in numbers) of all cases, they used the left antenna (Fig. 2.1, Suppl. Fig. 1 and Tab. 2.1a) (X^2-Test: n=599, X^2=59.6, df=1, p<0.0001).

Most contacts (40.4 %) were observed when the soliciting bee was situated on the left side (sector left) of the offering bee so that the right antenna was closer to the donor. In the position where the bees were facing each other almost directly (sector zero) 35.2 % of the contacts were observed. Only 24.5 % of the contacts were observed when the soliciting bees were situated on the right side (sector right) of the offering bee and the left antenna of the soliciting bee was closer to the mouthparts of the offering bee (Fig. 2.1, Suppl. Fig. 1 and Tab. 2.1b).

In order to find out whether the preference of the right antenna was associated with the angle of the bees' body axes, we divided the feeding positions into seven subsectors (three left subsectors, three right subsectors and the sector zero). The use of the right antenna over the left antenna was significant in every sector (Tab. 2.1b, Fig. 2.1 and Suppl. Fig. 1) (X^2-Test: sector right: n=148, X^2=7.8, df=1, p<0.005; sector left: n=241, X^2=34.4, df=1, p<0.0001; sector zero: n=210, X^2=86.1, df=1, p<0.005).

The preference of the right antenna in the subsectors was significant for all left sectors, the sector zero and the first subsector of sector right (sector right I) (for details see Suppl. Tab. 1).

Worker bees which had lost their first five segments of one flagellum still fed and got fed by other bees which were lacking these five segments as well.

If parts of one antenna were removed, the use of the opposite antenna increased in general (Fig. 2.2, 2.3, Suppl. Fig. 1 and Tab. 2.1).

Removing parts of the right antenna led to an increased use of the left antenna, but only up to 49 %. A smaller percentage (10.5 %) used the severed right antenna, and worker bees tried the amputated right antenna first and used the left intact one afterwards a relatively large number of times (40.5 %) (Fig. 2.1, Suppl. Fig. 1 and Tab. 2.1c) (X^2-Test: n=153, X^2=0.6, df=1, p<0.8).

If parts of the left antenna were removed, the use of the right antenna increased to 75 %. A smaller percentage (14.4 %) still used the severed left antenna, and in (10.3 %) of the feeding contacts, the workers first tried the stump and used the right, intact antenna afterwards (Fig. 2.1, Suppl. Fig. 1 and Tab. 2.1d) (X^2-Test: n=146, X^2=37.5, df=1, p<0.0001).

Such tendencies were recognizable in every sector (Tab. 2.1e - j, Fig. 2.1 and Suppl. Fig. 1) (X^2-Test: sector right e): n=50, X^2=0.8, df=1, p<0.8; sector right f): n=47, X^2=15.5, df=1, p<0.0001; sector left g): n=69, X^2=0.7, df=1, p<0.4; sector left h): n=56, X^2=20.6, df=1, p<0.0001; sector zero i): n=34, X^2=1.6, df=1, p<0.3; sector zero j): n=47, X =12.3, df=1, p<0.0001).

Comparing the use of the stump shows that the left stump was used significantly less often for trying a trophallactic contact before switching to the intact antenna in every sector (X^2-Test: all sectors: n=77, X^2=28.7, df=1, p<0.0001; sector right: n=21, X^2=13.8, df=1, p<0.0001; sector left: n=41, X^2=17.8, df=1, p<0.0001; sector zero: n=21, X^2=13.8, df=1, p<0.0001).

Coherence between the lopsided use of the antennae and an individual preference for one side could not be confirmed. The 30 individually marked workers that were observed showed the same preference for the right antenna as the bees in the first experiment did. The observed feeding activity of individual workers ranged between 3 and 11 feeding contacts, whereas only two bees were observed to use the right antenna exclusively in each of their three performed feeding contacts. Since there were only two cases in which honeybees used solely one side, there is no evidence for exclusive preference of right or left extremities (Fig. 2.4).

In addition, we found a significant difference in gustatory response scores (GRS) between stimulation delivered separately to the left and right antennae of individual bees. Right antennae were more sensitive for eliciting a sugar water induced PER than left antennae (Fig. 2.5) (Wilcoxon test (2-tailed): (R) n=52, (L) n=52, Z=-2.88, p<0.01). The score difference between right and left was positive and significantly different from zero, implying that individuals had predominantly more sensitive right antennae (Fig. 2.5) (one-sample t-test: n=52, mean=0.71, SD=1.86, SE=0.25, p<0.01).

2.5 Discussion

We observed a preference in soliciting honeybees for using the right over the left antenna for touching the mouthparts of the donor bee and consequently for receiving food from the donor (Fig. 2.1, Suppl. Fig. 1 and Tab. 2.1).

In addition, the preference of the right antenna was not a matter of individuality, since individual worker bees used the antennae alternately, however the right antenna was used more often than the left one (Fig. 2.4).

Furthermore, we compared the use of intact and amputated antennae in the different treatment groups in the neutral sector as well as all the sectors left and right. Worker bees that the first five segments of their right flagellum removed, tried to use their stump and then turned to use the intact left antenna more often, than workers that had parts of their left antenna removed, tried to use their left stump before using the right intact one (Fig. 2.2, Fig. 2.3, Suppl. Fig. 1 and Tab. 2.1).

There could have been various reasons for the preference of the soliciting bees for using the right antenna over the left.

Firstly, it could have been based on the spatial arrangement of the honeybees at the initiation of the trophallactic contact. If a soliciting bee stands on the left side of the donor (sector left) its right antenna is closer to the mouthparts of the donor, than if it stands on the right side of the donor (sector right). If both participants are facing each other more or less directly (sector zero) both antennae of the soliciting bees had virtually the same distance from the mouthparts of the donor. Consequently, the use of the right antenna should have increased in sector left and decreased in sector right as well as the use of the left antenna should have increased in sector right and decreased in sector left. That was not the case, since the preference for the right antenna occurred every position except for the sectors right II and III the soliciting bee took in reference to the donor bee (Fig. 2.1 and Suppl. Tab. 1).

Secondly, our experiments on gustatory response scores (GRS) of the honeybees showed that the right antenna was generally more sensitive for eliciting a sugar water induced PER than the left antenna (Fig. 2.5). While a difference in the number of pore plates on the antennae was correlated with olfactory lateralization in honeybees (LETZKUS ET AL., 2006), a significant difference in the relatively small number of taste hairs on the antennal tip which was stimulated in our experiments seems unlikely. Since the sucrose sensitivity of taste hairs on individual antennal tips is highly variable (HAUPT, 2004, HAUPT & KLEMT, 2005), the number of taste hairs would not be a reliable indicator for the sensitivity of an antenna. The estimation of a difference in sensitivity of the two antennae of an individual by physiological means is difficult since it requires the measurement of all spike responses of taste hairs in question, simultaneously if possible.

It is also possible that the difference in sensitivity arises in the dorsal lobe (HAUPT, 2007) or further still unidentified connections in the brain and suboesophageal ganglion that are involved in the release of the PER.

The stronger sensitivity we could show by GRS might also be the reason for the increased use of the right antenna in our behavioral experiments. Using the more sensitive antenna to feel for the mouthparts of the donors seems like a logical consequence. However, we only counted the actual antennal movement before an actual trophallactic contact. If touching the offered food with the left antenna produced no trophallactic contact, then a large number of this data might be missing in our experiments. Anyway, while analyzing the tapes, we did not recognize unsuccessful tries to touch the mouthparts of a donor with the left antennae when the antennae of both participants were intact. In addition, our experiments with the amputated antennae do not support such possibility.

Continuous use of the right antenna over the left even after amputation of the five distal flagellomeres suggests that the preference is not based on actual response to sensory stimulation since the gustatory sensilla which can trigger the PER were missing after the amputation to a large extent. Anyway, the worker bees tried to use the right antenna. If the honeybees would use the left and the right antenna equally often to start a trophallactic contact, the results of the amputation experiment should have shown no difference between the use of left and right antenna after amputation.

A third reason could be the involvement of a developmental or learning process in honeybees to preferentially use the right antenna for soliciting food from the donor bees. The fact that bees younger than 48h use both antennae simultaneously at the initiation of a trophallactic contact while older bees use just one antenna to touch the donor's mouthparts (MONTAGNER & PAIN, 1971) supports this idea.

We used winter bees exclusively in our behavioral experiments, which mean that the bees were a least six weeks old and had been able to establish a possible preference for one antenna in soliciting food over a rather long period of time.

The preference of the right antenna appears about 48h after eclosion (MONTAGNER & PAIN, 1971). Continuous feeding contacts and the stronger sensibility of the right antenna, which we could show in the GRS experiments, may condition the young honeybee to an increased use of the right antenna due to an increased sensibility for gustatory stimuli This hypothesis is sound because young bees have very high response thresholds and may not even respond to nectar collected by part of the foragers with a PER. Under these near-threshold conditions, the higher sensitivity of the right antenna would create a bias for PER release through the right antenna. The operant conditioning of antennal movements, which is side-specific (KISCH & HAUPT, 2009) could be a substrate for the acquired lateralization.

Our findings that worker bees often used the stump of the ablated right antenna to touch the mouthparts of the donor bee, before switching to the intact left antenna in almost any position, gives evidence that the use of the right antenna, although it is not sensitive enough to trigger a PER anymore, seems to be an effect from a developmental or learned process. Additional experiments with newly emerged bees with one antenna covered or coated with silicone, could help to clarify whether the preference of one antenna is innate or learned in any manner.

There is strong evidence for lateralization in the honeybee brain concerning olfactory (LETZKUS ET AL., 2006) and visual learning (LETZKUS ET AL., 2008). In addition, these studies showed that learning and discrimination of odors are mediated primarily by the right antennal pathway, leading to the idea that sensory inputs from the right half of the body in general are used preferentially while foraging or feeding. Our results support these findings since they involve a sensory input from the right antenna and are connected to the feeding behavior, but as discussed before, the actual reason for the preference of the right antenna in our experiments might be as well connected to sensory abilities of the antennae themselves. It would be interesting to see to what extent lateralization in learning experiments is present in freshly eclosed bees.

The honeybee´s preference of the right antennae in both naturally and experimentally induced PER shows lateralized behavior as it was shown before in invertebrates such as octopusis, spiders and fruitflies (ADES & RAMIRES, 2002; HEUTS & BRUNT, 2005; PASCUAL ET AL., 2004). Whether the behavior we observed can be associated with actual lateralization of the brain, behavioral lateralization, or rather with a learned or developmental process remains unclear. Further behavioral experiments with newly hatched bees and coated antennae as well as physiological analysis of the neural activity in the gustatory system might help to clarify the open questions.

2.6 Appendix – Figures and Tables

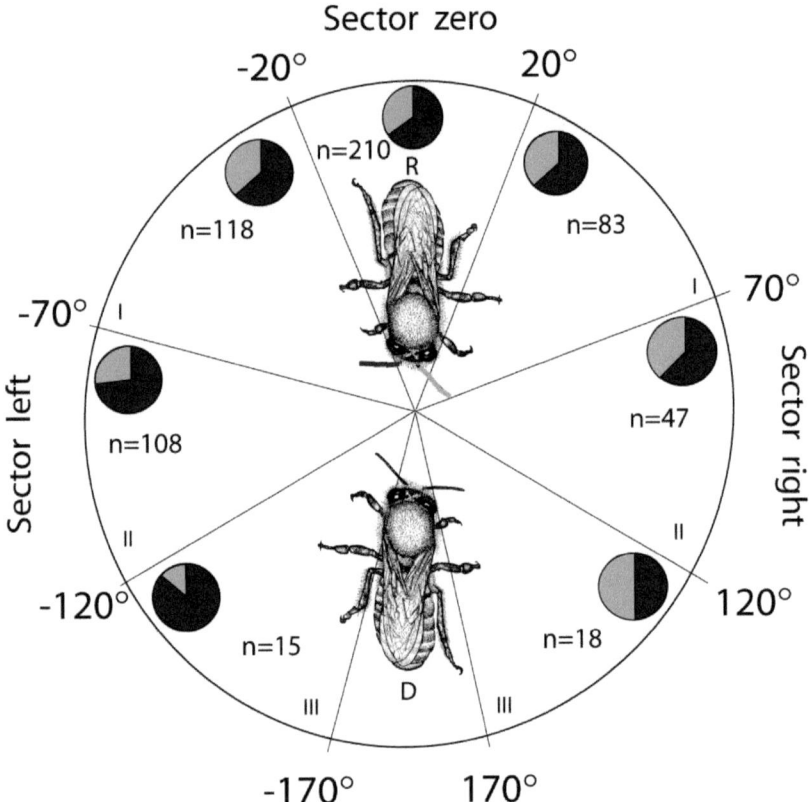

Fig 2.1 Feeding activities of recipient with both antennae intact

Pie chart showing the number of contacts (n) and the antennal use for positions where the recipient (R) was positioned left from the donor (D) (sector left), right from the donor (sector right) or was facing the donor (sector zero).
Sector left and right were additionally divided into three subsectors (I, II, III) each covering a 50° angle.
The use of the left antenna (light grey) and the use of the right antenna (drak grey) in the trophallactic contacts can be seen in each pie chart
The use of the right antenna (dark grey) dominated in almost every sector. Even in the subsectors of the right side, where the right antenna is at a greater distance from the mouthparts of the donor, the use of the right antenna is increased or at least equal to the use of the left antenna (For exact statstical analysis see Tab. 2.1 and Suppl. Tab. 1).

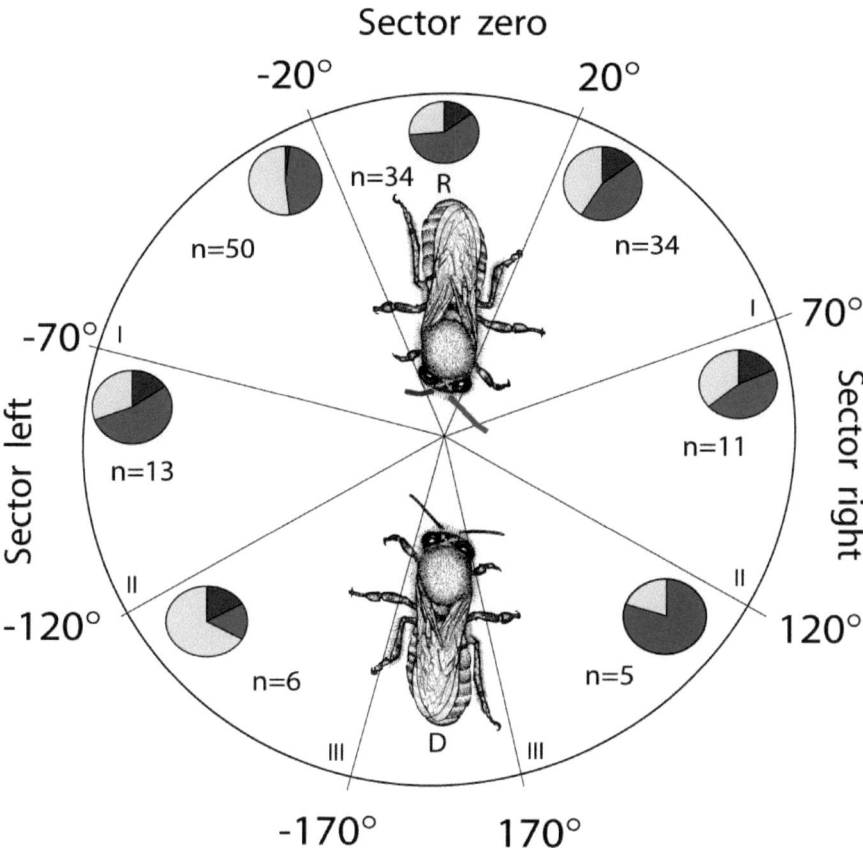

Fig 2.2 Feeding activities of recipient with right antenna amputated

Pie chart showing the number of contacts (n) and the antennal use for positions where the recipient (R) was positioned left from the donor (D) (sector left), right from the donor (sector right) or was facing the donor (sector zero).
The use of the left antenna (grey), the use of the right antenna (dark grey) and the use of the stump before switching to the left intact antenna (light grey) in the trophallactic contacts can be seen in each pie chart. The bees often tried the stump first and switched then to the intact left antenna and even used the stump (dark grey) exclusively to trigger a feeding contact in almost every sector. (For exact statistical analysis see Tab. 2.1 and Suppl. Tab. 1).

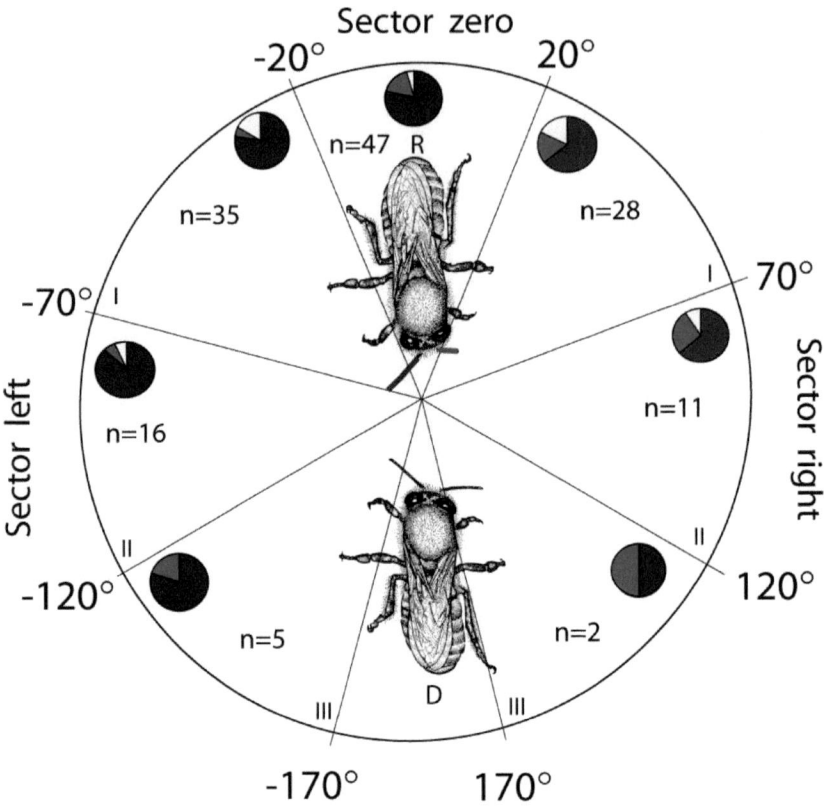

Fig 2.3 Feeding activities of recipient with left antenna amputated

Pie chart showing the number of contacts (n) and the antennal use for positions where the
 recipient (R) was positioned left from the donor (D) (sector left), right from the donor (sector right) or was facing the donor (sector zero).
The use of the left antenna (grey), the use of the right antenna (dark grey) and the use of the stump
before switching to the right intact antenna (white) in the trophallactic
contacts can be seen in each pie chart. The bees tried the left stump (white) less frequently than right stump (light grey) in Fig 2.2 before switching to the intact right antenna (dark grey). The left stump was used exclusively (grey) to trigger several trophallactic contacts in every subsector.
 (For exact statistical analysis see Tab. 2.1 and Suppl. Tab. 1).

Antennal dexterity in honeybees

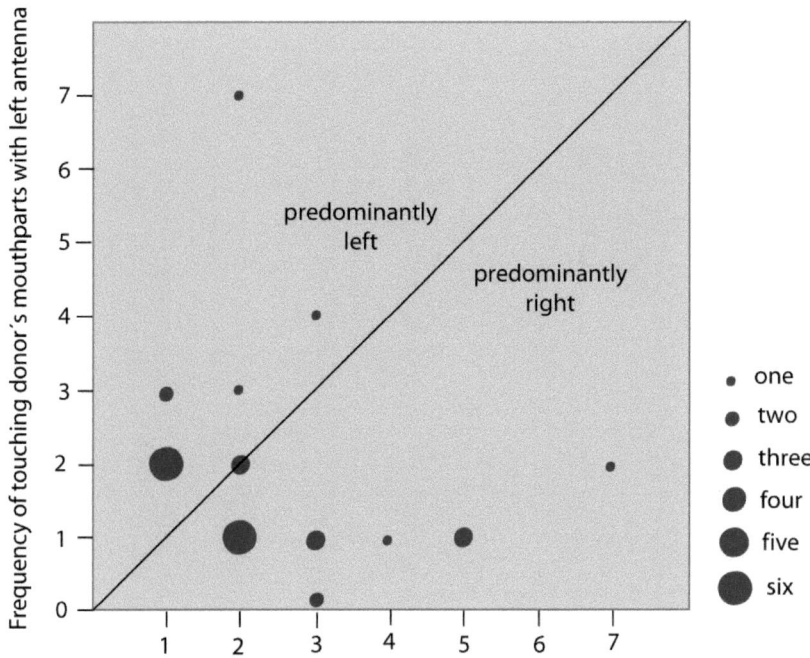

Fig 2.4 Test for individual preference of one antenna

Each spot reseambles one to six individuals (displayed by the size of the dots) and their balance between using the left and the right antenna (if both sides were used equally often the dot is on the black line). The feeding activity of an individual bee can be calculated by adding the value on the y- and the value on the x-axis.
If honeybees were using their antennae equally, the dots would be lined up along the diagonale, if they would use one side exclusively the dots were lined up on the x- respectively on the y-axis.

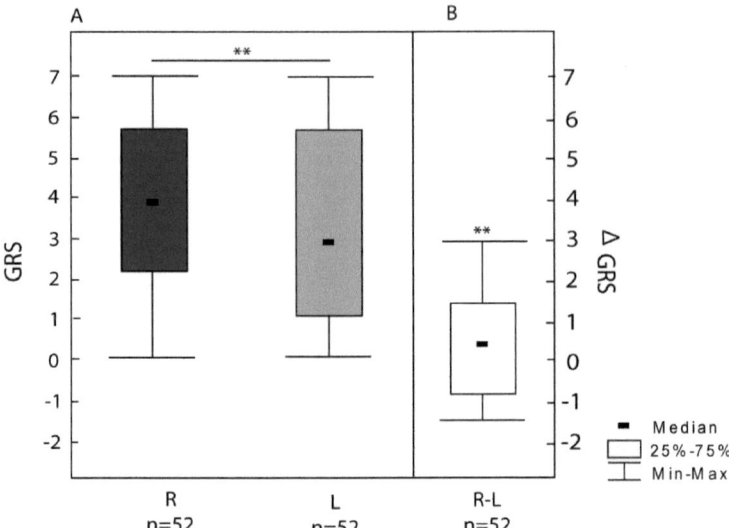

Fig. 2.5 Gustatory response scores (GRS)

A) Scores between the left and the right antenna were compared with the Wilcoxon test (2-tailed): $p<0.01$, $Z=-2.88$; $n(R)=52$, median=3.75, Q1=2.25, Q3=5.75, Min=0, Max=7; $n(L)=52$, median=2.5, Q1=1, Q3=5.75, Min=0, Max=7

B) The score differences were tested against the null hypothesis of being zero using the one-sample t-test: $n(R-L)=52$, mean=0.71, SD=1.86, SE=0.26; median=1.0, Q1=0.25, Q3=1.5, Min=0, Max=3

Suppl. Fig. 1

A)
Participation of left (red) and right (blue) antenna and the respective stump (yellow/green) of the soliciting bees for all sectors combined in percent (absolute numbers on the bars). (For details and statistics see Tab. 2.1)

B)
Sector zero, left and right sector itemized
Participation of left (red) and right (blue) antenna and the respective stump (yellow/green) of the soliciting bees for all sectors combined in percent (absolute numbers on the bars).
(For details and statistics see Tab. 2.1)

Right antenna

Left antenna

Tried stump first then switched to intact antenna

** indicates p< 0.01
*** indicates p< 0.001
n.s. = not significant

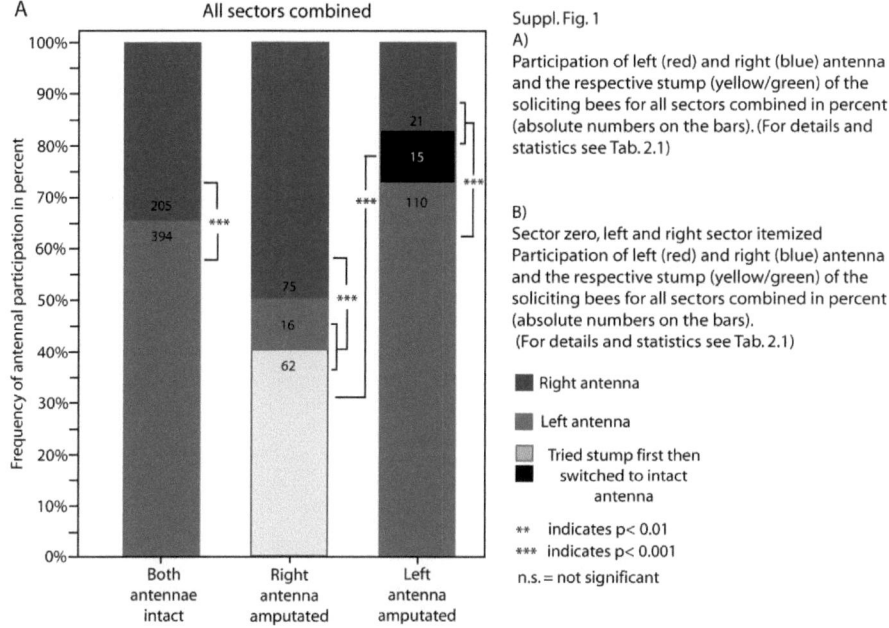

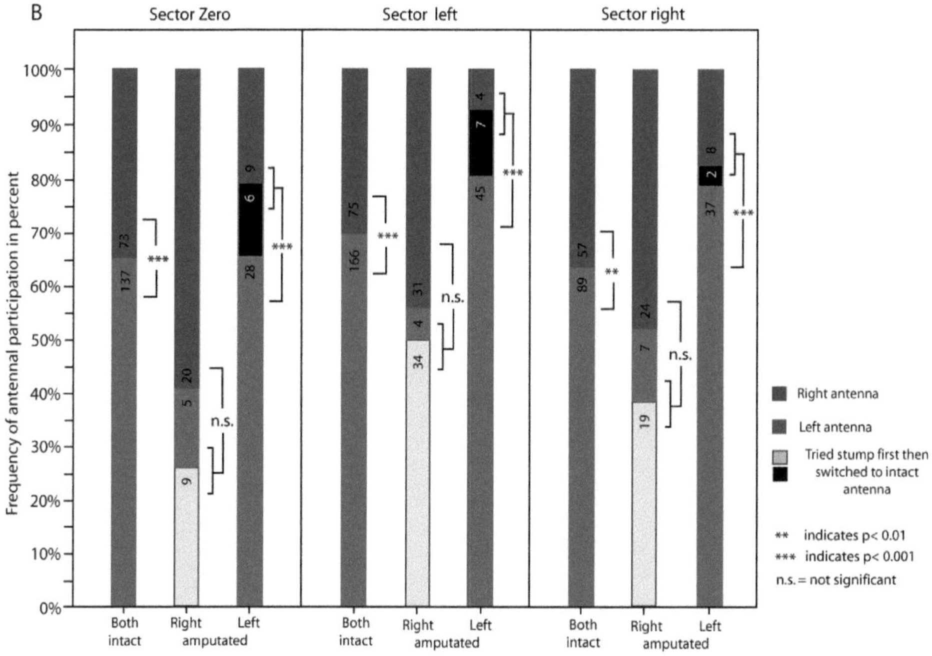

Antennal dexterity in honeybees

	Status of antenna	n	Right antenna	Left antenna	Tried stump first then used intact antenna	X^2 Test: Comparison of right & left
All sectors	a) Both intact	599	394 / (66 %)	205 / (34 %)	n.a.	$p<0.0001$ $X^2=59.6$
	c) Right ablated	153	16 / (10.5 %)	75 / (49 %)	62 / (40.5 %)	$p<0.8$ (n.s.) $X^2=0.6$
	d) Left ablated	146	110 / (75 %)	21 / (15%)	15 / (10%)	$p<0.0001$ $X^2=37.5$
Sector right	b) Both intact	148	91 / (61 %)	57 / (39 %)	n.a.	$p<0.005$ $X^2=7.8$
	e) Right ablated	50	7 / (14 %)	24 / (48 %)	19 / (38 %)	$p<0.8$ (n.s.) $X^2=0.8$
	f) Left ablated	47	37 / (79 %)	8 / (17 %)	2 / (4 %)	$p<0.0001$ $X^2=15.5$
Sector left	b) Both intact	241	166 / (69 %)	75 / (31 %)	n.a.	$p<0.0001$ $X^2=34.4$
	g) Right ablated	69	4 / (6 %)	31 / (45 %)	34 / (49 %)	$p<0.4$ (n.s.) $X^2=0.7$
	h) Left ablated	56	45 / (80.5 %)	4 / (7 %)	7 / (12.5 %)	$p<0.0001$ $X^2=20.6$
Sector zero	b) Both intact	210	137 / (65 %)	73 / (35 %)	n.a.	$p<0.0001$ $X^2=86.1$
	i) Right ablated	34	5 / (14.5 %)	20 / (59 %)	9 / (26.5 %)	$p<0.3$ (n.s.) $X^2=1.6$
	j) Left ablated	47	33 / (76 %)	8 / (18.5 %)	2 / (4.5 %)	$p<0.0001$ $X^2=12.3$

Tab 2.1 Antennal uses in trophallactic activity

The antennal uses during trophallactic contacts in every sector compared with an X^2-test. The value of "Tried stump first then used intact antenna" were added to the ablated antenna (df=1 in all cases).

Antennal dexterity in honeybees

Sector	n	X^2	df	p
Sector left I	15	8.1	1	p<0.005
Sector left II	108	23.1	1	p<0.0001
Sector left III	118	7.63	1	p<0.006
Sector zero	210	19.5	1	p<0.0001
Sector right I	83	5.31	1	p<0.02
Sector right II	47	2.57	1	p<0.1 (n.s.)
Sector right III	18	0	1	p=1 (n.s.)

Suppl. Tab. 1 Antennal use in trophallactic activity
Differences (analyzed by X^2-test) between the use of left and right antenna in trophallactic activity per subsector with both antennae intact.

3. Does sugar equal heat? – Sugar intake and its impact on thoracic heat production in the honeybee

3.1 Abstract

Honeybees are heterothermic insects which can actively regulate their thoracic temperature by shivering their flight muscles. Their heat capacity is influenced by parameters as behavior and food quality. Even though the individual can regulate its body temperature, its heating performance is strictly limited by the honey ingested. The reason for this is that honeybees use mostly the glucose in their hemolymph as energy substrate for muscular activity and the heat producing flight muscles are among the metabolically most active tissues known.

The fuel for their activity is honey; which is processed nectar with a sugar content of ~80 % stored in the honeycomb. Freshly collected nectar has a relatively high water content which must evaporate to ripen the honey before capping. Conversely, they have to dilute the stored honey before they can metabolize it later.

We found that the sugar content of the ingested food affects the thoracic temperature of the honeybees even if they show no heating-related behavior. Caged honeybees displayed high thoracic temperatures with increasing sugar content up to 65 % sugar concentration. In addition, a water supply seems to be of particular importance for the utilization of sugar in thermogenesis. Water is a basic ingredient in the metabolic pathway of energy gain in the honeybee's flight muscles and there is evidence of severe loss of hemolymph and therefore water while flying and heating.

Currently, there is little knowledge about the sugar/water equilibrium in honeybee heating physiology and the water needs of the individual honeybee in general. Testing the optimal sugar/water equilibrium and providing new information about it as a factor for physiological condition and nutritional needs might support research on malnutrition in honeybees.

The impact of sugar on the thoracic temperature in honeybees and other hymenopterans has been discussed in connection with certain behaviors in several studies. The elevated body temperature of workers after feeding on food with high sugar content was even suspected to provide information about the food source to the recipient of a trophallactic contact. At any rate, there have been no studies to test the direct impact of sugar content on thoracic temperature so far.

3.2 Introduction

Honeybees are considered as partially homoeothermic (HOFFMANN, 1978) if one regards the colony as a superorganism. They only reach ambient temperature when inactive or resting, but are capable of raising their thoracic temperature before going on a foraging flight, for heating the brood or warming the core of the winter cluster. The ability of honeybee workers to generate large amounts of heat through so called "shivering thermogenesis" (STABENTHEINER ET AL., 2003) depends to a large extent on the glycogen metabolism (PANZENBÖCK & CRAILSHEIM, 1997).

Honeybees use mostly sugar as an energy substrate for muscular activity (JONGBLOED & WIERSMA, 1934; LOH & HERAN, 1970; SACKTOR, 1970; ROTHE & NACHTIGALL, 1989) and the level of glycogen in the hemolymph must be kept high to provide an adequate fuel supply for the heat-generating flight muscles (CRAILSHEIM, 1988) which are the most metabolically active tissues known (SOUTHWICK & HELDMAIER, 1987). There is no alternative to heat production by shivering thermogenesis in the honeybee. The species *Bombus* e.g. can produce heat alternatively by metabolizing glucose with the aid of the enzyme fructose-1.6-diphosphatase. Honeybees are unable to metabolize glucose in this alternate way, since the activity of the required enzyme is rather low in the honeybee flight muscle (less 0.05 µmol per g) (NEWSHOLME ET AL., 1972).

Honeybees generally remain endothermic as long as they have sugar in their honey stomach or midgut. When the food is consumed, they soon exhaust their tissue reserves and die. A crop load of sugar solution can provide a bee with food for several hours. But even inactive, caged honeybees with a full crop held at room temperature die within 7h after being separated from their food source (HEINRICH, 1993).

SOTAVALTA (1954) reported that honeybees he kept flying for 10 to 15min died 5 to 10min later unless food was given to them. They normally have energy supplies in the form of honey from the comb or food from the crops of their nestmates nearly constantly within reach, and so they do relatively little to conserve them (HEINRICH, 1993).

Does sugar equal heat?

Young bees only gradually develop the capacity for endothermic heat production (HIMMER, 1932; ALLEN, 1955; HARRISON, 1986; STABENTHEINER & SCHMARANZER, 1987). Before they have developed the capacity to generate heat by shivering, new workers tend to stay in the warm brood nest (FREE, 1961). Within the first few days, the maximal thorax-specific metabolic rate closely corresponds to the increase in enzyme activities. Pyruvate kinase and citrate synthetase activities increase (tenfold) up to only 4 days of age, and then gradually decline (HARRISON, 1986).

By contrast, BUJOK (2005) demonstrated that bees show proper brood heating activity 48h after eclosion, even though their physiology should not be fully adapted to this task. Since young bees kept in cages outside the hive and without direct access to a queen show signs of higher JH activity which is known to have a potent effect on muscle growth, the flight capability (WYATT & DAVEY, 1996) and the respiratory metabolism (NOVAK, 1966) in insects, both findings are not mutually exclusive.

The basic food requirements of the worker honeybee include mostly carbohydrates in the form of sugar and water to dilute it (LINDAUER, 1955; HAYDAK, 1970). Freshly collected nectar contains 25 to 75 % sugars in dry matter (BAKER & BAKER, 1983), and as such is particularly susceptible to bacterial degradation. Bees have evolved a suite of adaptations to be able to survive year-round on this source of food. These adaptations involve the conversion of nectar into honey, with physical and chemical properties that contribute to its long-term stability. By evaporating moisture from the nectar they convert nectar into honey, a supersaturated solution, with high osmotic potential, making it difficult for bacteria to survive in it (CRANE, 1996). Nectar is converted to honey not only by evaporation of water, but by three enzymes secreted by the hypopharyngeal glands of workers. Alphaglucosidase converts sucrose, the primary component of nectar, into glucose and fructose (SIMPSON ET AL., 1968; KUBO ET AL., 1996; OHASHI ET AL., 1996, 1997). Amylase hydrolyses plant starches that contaminate the nectar (WINSTON, 1987; OHASHI ET AL., 1999). Finally, a glucose-oxidase converts glucose into gluconicacid and peroxide, both of which afford antiseptic activity to the honey (WHITE ET AL., 1963; OHASHI ET AL., 1999; KUNIEDA ET AL., 2006).

When feeding, the bees dilute the honey with water either from outside the hive, or from their own metabolism (ALTMANN, 1956; ALTMANN & GONTARSKI, 1963). It is a well known fact that a colony of bees utilizes large amounts of water to dilute honey and to regulate temperature in the brood nest (LINDAUER, 1955). During the spring and early summer months, bees collect large quantities of water for use in the hive for such purposes as softening down winter stores, etc. (BUTLER, 1940). SEELEY (1995) estimated average annual requirements of 25 liters of water for a single wild colony.

The individual water economy of the bees is influenced by hormones secreted from the *corpora allata* and *corpora cardiaca*, the first increasing water consumption, the latter decreasing it (ALTMANN, 1953). The presence of these hormones is coupled with season and ambient temperatures. If temperatures drop to less than 20 °C as in the cold season, the hormone from the corpora cardiaca increases the water permeability of the midgut and hindgut, withholding more water in the tissue and reducing the feces volume in the rectum. This condition simplifies the lack of water in the winter hive by increasing the efficiency of the use of the body's water resources and increases the time between defecation flights in winter.

Worker bees lose more water due to the lower permeability of their digestive tract in the warmer season, but they have virtually constant access to water collected by foragers and are usually not prevented from making defecation flights by ambient temperatures lower than 10 °C.

While in vertebrates only a minor part of their body water is contained in the blood (e.g. fish 2.7 % and dog 5.4 %), the honeybee's blood contains 25 to 30 % of the whole water proportion of the individual. This indicates that the honeybee's blood plays an important role for its individual water balance (HOFFMAN, 1978). Water is a basic ingredient in the metabolic pathway of energy gain in the honeybees' flight muscle. Adenosine triphosphate hydrolysis and its regeneration in the insects' working flight muscle require a sufficient water supply (WEGENER, 1996) from the hemolymph. Even though there is high metabolic water production during flight in hymenopterans (BERTSCH, 1984; NICOLSON & LOUW, 1982), there is also evidence of severe loss of hemolymph while flying. ALTMANN (1956) found that the water loss of a flying bee is very high and it can suffer the loss of nearly all the water in its hemolymph.

Brood heating and flying rapidly consume the energy reserves of a worker bee and can be referred to as physiologically equivalent, since the same muscles are active (HEINRICH, 1993).

The hot thorax itself is another source of water loss. Isolated bees lose a large amount of water through the mouth and the labrum. This effect increases if the bees are confronted with a CO_2 load, for example when they crowd together either on the brood or in the winter cluster. Heating bees exhale CO_2, the resulting waste-product of the respiratory chain. The accumulation of CO_2 in the cluster forces the bees to open their spiracles wider (LOUW & HADLEY, 1985) and accordingly they lose a higher amount of water by evaporation.

These two factors, the use of water in ATP hydrolysis and the indirect loss of water due to CO_2 in the environment, emphasize the particular importance for the bees of keeping the water balance to maintain heating ability.

Even though a honeybee colony needs large amounts of water to dilute the stored honey, there is no actual data concerning the amount needed to dilute honey for optimum heating results. Since the flight muscles produce large amounts of metabolic water which evaporates during flight to a large extent, we expect that the metabolic water from heating activity might have a major impact on heat production.

The modulation of the thoracic temperature of the honeybee can be explained by the effect of ambient temperature and/or the behavioral situations. Flying increases the thoracic temperature as a by-product of muscular activity in every insect. Accordingly elevated thoracic temperature in honeybees can be observed while flying. In several species of the genus *Apis* it has been demonstrated that sugar concentration and foraging distance are other factors which modulate the thermal behavior of honeybees (DYER & SEELEY, 1987; STABENTHEINER & SCHMARANZER, 1988; SCHMARANZER & STABENTHEINER, 1988; WADDINGTON, 1990; STABENTHEINER & HAGMÜLLER, 1991; UNDERWOOD, 1991; STABENTHEINER ET AL., 1995; STABENTHEINER, 1996). At the feeding place, the thoracic temperature changes according to the food quality (SCHMARANZER & STABENTHEINER, 1988), and after returning to the hive, the thoracic temperature even while dancing, walking and trophallaxis is positively correlated with the sugar content of the food (STABENTHEINER & HAGMÜLLER, 1991). The correlation between sugar concentration at a feeder and the trophallactic behavior when unloading the nectar has even been conjectured as exchange of information about the food source between forager and nectar recipient (FARINA & WAINSELBOIM, 2001B).

There is evidence for a direct impact of the sugar content on the thoracic temperature in other hymenopterans. NIEH ET AL. (2006) showed a positive correlation between thoracic temperature and sugar concentration in *Bombus wilmattae* and NIEH and SANCHEZ (2005) a similar effect in *Mellipona panamica*. KOVAC and STABENTHEINER (1999) observed the influence of sugar concentration on thoracic temperatures in *Vespula vulgaris* at an artificial food source and presented similar results. However, the distance between a food source and nest could not be precluded as a parameter for elevated temperatures in these experiments.

The fact that the positive correlation between the thoracic temperature and the sugar content of the food is present at different behaviors and in different hymenopterans indicates that the sugar content might have a direct influence on the individual's thoracic temperature, not only in certain species and at certain behaviors, but in general. Since all previous experiments and observations are connected to behaviors (foraging, dancing etc.) which are also influenced by various biotic and abiotic factors (ambient temperature, barometric pressure and air humidity or the distance between hive and food source etc), a clear conclusion how sugar itself affects the thoracic heat production could not be drawn until now. We expect that the sugar intake and the water supply have a direct impact on the thoracic temperature of the honeybee even under laboratory conditions and in a task free or "neutral" environment.

3.3 Materials and Methods

All observations were made at the Bee Station of Würzburg University (Biocenter) from May 2007 to July 2007 and November 2007 to January 2008 with *Apis mellifera carnica* in a shaded and climate controlled room (20 °C ± 1) under red light conditions.

Workers were used from four different colonies headed by unrelated queens. In order to provide genetic variance, each of the queens was artificially inseminated with the sperm of 12 different drones.

The bees were kept in small wooden boxes (7x11x5cm limited by a grating at the rear and covered with heat radiation permeable foil at the front side. We used the plastic wrap "Cling Wrap"© made of 100 % polyethylene manufactured by the U.S. American company GLAD™. Since the foil still alters the radiation averting from the object and the radiation measured by the camera, we needed to define the error produced at different temperatures. We equipped the thoraces of dead bees with small carbon film resistors which we connected to a transformer (Amrel Linear Power Supply LPS-301) with an output of a constant voltage warming up the resistor. Changing the output of the transformer changes the heat emitted from the resistor, enabling us to measure the temperature emitted by the thoraces between 20 °C and 42 °C with and without the foil. The error produced by the foil is non-linear (Suppl. Fig. 2); therefore every thoracic temperature measured in the experiment had to be specifically corrected (Suppl. Tab. 2).

Each box contained 10 worker bees and artificial nectar at a different sugar concentration. All bees were able to move freely in the box and had open ad libitum access to the food. As food we used Apiivert©, commercially available sugar syrup which is most common for beekeeping in Germany, consisting of 72.7 % dry matter, whereof 39 % is fructose, 31 % dextrose and 30 % sucrose. For our experiments we diluted the sugar syrup with water until it contained the required concentration of sugar. In our summer experiments we measured a sugar concentration of 75 % in Apiinvert; in winter we measured 72 % sugar content. We used the undiluted syrup at its highest concentration and diluted it with water in steps of 5 and 10 %. Each solution was prepared fresh and checked with a hand held refractometer before feeding it to the honeybees. Other researchers often measure the sugar content in mol; therefore we provide a table for easy conversion of the values (Suppl. Tab. 2).

Thermal images of the thoracic temperatures of the honeybees were recorded with a thermal imaging camera S40 (FLIR Systems Inc.). We took a still picture of the boxes and analyzed the thoracic temperatures of all bees every minute. Honeybees are able to heat up or cool down relatively quickly (approximately Δ 10 ℃ per min); therefore we chose steps of one minute between the still pictures giving each bee the opportunity to change its temperature. Each sugar concentration was tested 3 times in every experiment. To ensure that the bees were metabolizing solely the defined sugar syrup, every treatment group was kept for two days with their specific solution before measuring. The solution was replaced every day.

The bees can only be identified if their temperature is different from the ambient temperature. An inactive bee which does not produce heat or is not passively heated up by other bees has automatically ambient temperature. We took care that all bees were in a proper physiological state and able to produce heat before the shots, otherwise a dead bee would have been counted as "non heating" and with ambient temperature for the wrong reason.

3.3.1 Set up without additional water

We chose 11 different sugar concentrations from 25 % to 75 % content in steps of Δ 5 % sugar content from May 2007 to July 2007. The bees were collected as red eyed pupae (brood comb) and allowed to eclose on their combs which had been placed in an incubator (35 °C). After 48h, 10 bees were placed in each box, and fed with one specific sugar concentration. After another 48h the boxes were packed closely so the camera could take shots from all groups at the same time for two consecutive hours.

3.3.2 Set up with additional water

We conducted experiments with 6 different sugar concentrations between 22 % and 72 % in steps of 10 % sugar content and additional water supply in the box. The bees were taken from one of four colonies for every run from November 2007 until January 2008. Since there had been no brood in these hives since October 2007 the bees must have been at least four weeks old. For one experimental run, 10 bees were placed in each box, and fed with one specific sugar concentration. After 48h the boxes were packed closely so the camera could take shots from all groups at the same time for two consecutive hours.

All statistical analyses were performed using the statistical package Statistica 8©.

3.4 Results

3.4.1 Set up without additional water

The thoracic temperatures varied between the 11 differently fed groups. Bees receiving sugar concentrations of 45 % and 65 % achieved highest median temperatures. Bees that received food with low sugar concentrations (25 and 30 %) had lower median temperatures than the other groups. Sugar concentrations of 35, 40, 50, 55, 60, 70 and 75 % made bees produce average median thoracic temperatures (Tab. 3.1, Fig. 3.1).

Maximal temperatures were rather steady (Tab 3.1). Even in groups fed with low sugar syrup, maximal temperatures of at least 33.6 °C were reached. Only in the groups with the highest median temperature, the maximum temperature was elevated as well (Tab. 3.1).

The thoracic temperatures of all groups differed significantly, except for the pairs 35 and 70 %, as well as 50 and 55 % (Fig. 3.1). (Kruskal-Wallis Anova: n=39319, H=9320, p<0.0001. For multiple comparisons of all groups and exact p-values see Suppl. Tab. 3).

Even though the median temperatures did not increase in step with sugar content, the sum of the data showed a significant positive correlation between sugar content and thoracic heat production (Spearman rank correlation: n=40309, R=0.33, p<0.05).

Does sugar equal heat?

If counting only the hottest bee per picture, the main results repeat to a large extent: bees that received food with the lowest sugar concentrations had lowest median temperatures. Bees receiving sugar concentrations of 45, 60 and 65 % achieved highest temperatures. The thoracic temperatures of all groups differed significantly, except for the pairs 25 and 70 %, 35 and 40 %, 35 and 70 %, 45 and 60 %, 50 and 55 %, 50 and 75 % as well as 55 and 75 % (Fig. 3.2, Tab. 3.2). (Kruskal-Wallis Anova: n=4031, H=1173.5, p<0.001. For multiple comparisons of all groups and exact p-values see Suppl. Tab. 4).

The sugar content of the food and the thoracic temperatures of the hottest bee per group and picture were positively correlated as well (Spearman rank correlation: n=1320, R=0.32, p<0.05).

3.4.2 Set up with additional water

The thoracic temperatures varied between the 6 differently fed groups. Bees receiving sugar concentrations of 62 and 72 % achieved highest thoracic temperatures. Bees that received sugar with a concentration of 22 % had lower mediated temperatures than the other groups. Minimal temperatures were near room temperature and showed very little variance (Δ 0.1 °C). Even groups fed with lowest sugar concentrations reached equally high maximal thoracic temperatures as did the other groups and the maximal temperatures showed relatively little variance as well (Δ 1.4 °C). The thoracic temperatures of all groups differed significantly, except for 62 and 72 %. (Fig. 3.3, Tab. 3.3) (Kruskal-Wallis Anova: n=21780, H=3033.3, p<0.0001. For multiple comparisons of all groups and exact p-values see Suppl. Tab. 5).

The thoracic temperatures increased stepwise with the increasing sugar concentration which led to a significant positive correlation between thoracic heat production and sugar content of the syrup (Spearman rank correlation: n=21780, R=0.35, p<0.05).

If counting only the hottest bee per picture, the main results repeat to a large extent: bees that received the lowest sugar concentrations (22 %) had lowest median temperatures. Bees receiving nectar concentrations of 62 % and 72 % achieved highest temperatures. The thoracic temperatures of most groups differed significantly, except for 32 and 42%, 32 and 52 %, 42 and 52 %, as well as 62 and 72 % (Fig. 3.4, Tab. 3.4). (Kruskal-Wallis Anova: n=2178, H=401.7, p<0.0001, For multiple comparisons of all groups and exact p-values see Suppl. Tab. 5).

There was a positive correlation between the sugar content and the thoracic temperatures of the hottest bee per group and picture as well (Spearman rank correlation: n=2178, R=0.39, p<0.05).

3.5 Discussion

The thoracic temperature of the bees showed a clear dependence on the sugar content of the sugar they consumed. Low sugar content nectar always resulted in lowest median temperatures in our experiments. Accordingly, the sugar intake had a direct impact on the thoracic temperature, even though the increase in thoracic temperatures differed among the treatment groups subject to the availability of additional water.

While workers with an additional water supply showed increasing median thoracic temperatures with every increase in sugar content, the increase of thoracic temperatures in bees without additional water supply was not linear. However, there were significant positive correlations between sugar content and thoracic temperatures in both groups. Therefore our data supports previous findings of STABENTHEINER and SCHMARANZER (1988); DYER and SEELEY (1987); SCHMARANZER and STABENTHEINER (1988); WADDINGTON (1990); STABENTHEINER and HAGMÜLLER (1991); UNDERWOOD (1991); STABENTHEINER ET AL. (1995); STABENTHEINER (1996) which all found a positive correlation between sugar content of the food and the honeybees´ thoracic temperature at certain tasks or while executing certain behaviors. In our experiment "behavior" was excluded as a variable to a large extent, by keeping the bees in a neutral or "task free" environment, where thoracic temperatures are not affected by flight, brood incubation, or low ambient temperatures. Previous experiments on the change in thoracic temperature and behavior according to variations in sugar content regarded the increase in thoracic temperature as a behavioral consequence of a high quality food source (DYER & SEELEY, 1987; STABENTHEINER & SCHMARANZER, 1988; SCHMARANZER & STABENTHEINER, 1988; WADDINGTON, 1990; STABENTHEINER & HAGMÜLLER, 1991; UNDERWOOD, 1991; STABENTHEINER ET AL., 1995; STABENTHEINER, 1996) and were even suspecting that an increase in thoracic temperature due to the higher sugar content of the food source might have an informational value for recruiting other foragers (FARINA & WAINSELBOIM, 2001B). Our results do not confirm such interrelation. Our experiments show that an increase in thoracic temperature was rather a consequence of the food´s sugar content or more precisely a consequence of a high blood sugar level in general.

Does sugar equal heat?

Honeybees' blood sugar is known to influence the thoracic temperature through the activity of the muscle directly (JONGBLOED & WIERSMA, 1934; LOH & HERAN, 1970; SACKTOR, 1970; ROTHE & NACHTIGALL, 1989) and the blood sugar in turn is extensively influenced by the sugar content of the food. CRAILSHEIM, (1988) and ABOU-SEIF ET AL. (1993) found positive correlations between the sugar solution they fed bees and the measured hemolymph sugar levels. The reason for this connection seems to be that there is no hemolymph sugar homeostasis (CANDY ET AL., 1997). Therefore low sugar content nectar will accordingly lead to low blood sugar which entails less efficient muscle work and consequently will end in lower thoracic temperatures.

BLATT and ROCES (2001) stated that the variability in hemolymph sugar levels in honeybees might be a side effect of different experimental conditions, causing different levels of activity and resulting in metabolic differences. Our data suggest that their objection in consideration of differences due to different levels of activity was justified. Several honeybees within the groups fed with lowest sugar concentrations reached relatively high thoracic temperatures. These exceptionally hot bees must have found a way to raise their blood sugar and their thoracic temperature against the low sugar content of the provided food. The reason for the unexpected temperature increase might be an effect created by trehalose synthesis, an alternative metabolic pathway which is known to stabilize the rise in the blood sugar in honeybees fed with nectar containing 30 and 50 % sugar (BLATT and ROCES, 2001). Nevertheless, this alternative seems to be the exception, since the median temperatures of the groups fed on low sugar content food remained far below those of the other treatment groups.

Another explanation for the elevated thoracic temperatures in several bees despite low sugar content food, is the concentration of nectar by evaporation. Honeybees are known to concentrate nectar by evaporating the water contained to a large extent (PARK 1925; NICOLSON & HUMAN, 2008). Even though the nectar was replaced every day to prevent evaporation, bees are capable of concentrating the nectar they ingest. Therefore, some bees might have been able to raise their blood sugar by altering the dosage of sugar in the nectar they were provided.

Minimal temperatures in our experiments were rather equally distributed since the laboratory was climate controlled to 20 °C ± 1 and inactive bees always showed ambient temperature.

We can confirm that the coldest workers were were inactive, since the thoracic heat production which involves activity would have caused a visible contrast between thorax and head thorax and abdomen respectively in the thermal imaging shot.

Nonetheless, there are certain variations in minimal thoracic temperatures in worker bees without additional water. In some groups several workers showed minimal thoracic temperatures which were Δ 1 °C higher than actual room temperature. This might be an effect of the elevated median temperatures in these groups. Because the cages were rather small and covered by plastic foil, heat accumulation might have enhanced and consequently elevated the ambient temperature in these specific cages. This theory is corroborated by the fact that highest minimal temperatures were achieved in summer bees fed on 65 % sugar solution, which also have highest maximal and highest mediated thoracic temperatures.

Some of the groups showed high variances (Q1-Q3) in thoracic heat production even though the 10 worker bees were treated equally. Worker bees fed on 35 % sugar solution without additional water supply had a variance of Δ 7.5 °C in thoracic temperature (hottest bee per picture 35 % sugar content: Δ 8.1 °C variance). This wide variety might be an effect of the already described abilities of the honeybees to concentrate the sugar provided with the nectar. Worker bees with additional water supply show similar variances at sugar concentrations of 32 % with a variety of Δ 8.5 °C (hottest bee per picture 32 % sugar content: Δ 10.9 °C variance). In addition, worker bees with additional water supply showed strong variances in thoracic temperature at nectar concentrations with 72 % sugar content (Δ 8.4 °C variance). This wide variety might be an effect of the different abilities of the honeybees to dilute the nectar with the water provided in the box. Since there has been no additional behavioral study concerning the actual intake of food or water, this assumption must remain unverified.

The sugar solution which caused highest thoracic temperatures in both experiments was a concentration of 65 % (in the set up without additional water) and 62 and 72 % (in the set up with additional water) which seems to be the optimum sugar/water equilibrium for heating. DETROY ET AL. (1981) published results about the food requirement of caged honeybees and considered 67 % sugar syrup as the optimum, because they lost a lower number of bees kept at this sugar concentration. Interestingly, this optimum sugar concentration is rather close to our sugar concentration 65 % respectively 62 and 72 % which was metabolized best in both experiments.

Does sugar equal heat?

Syrups of higher sugar content were metabolized equally well in workers given additional water. In workers without additional water, higher concentrations led to lower median thoracic temperatures. This might have been an effect of the low water content of the food, or the higher thoracic heat produced by higher sugar content led to a higher rate of evaporation and therefore reducing the water content to a level which was intolerable for high metabolic rates. In addition, the increased thoracic temperature could be related to higher CO_2 production which stimulates the bees to open their spiracles, thereby increasing evaporation of water and contributing to the water loss of individual bees (LOUW & HADLEY, 1985).

Furthermore, worker bees without additional water supply showed relatively high thoracic temperatures while being fed with 45 % sugar solution.

ALTMANN and GONTARSKI (1963) found that the sugar concentration in the crop of a bee is usually lower than that of the previously ingested food. They tested two different sugar solutions with 54.1 % (± 1 %) and 80.2 % (± 1.7 %) sugar content without an additional water supply. The measured crop contents of the bees were 48.0 % (± 3.4 %) and 67.5 % (± 4.2 %) and therefore lower in sugar content. So the syrup must have been diluted by the honeybees. Interestingly, these values match the sugar concentrations that were metabolized best by the worker bees without additional water in our experiments. If the worker bees dilute their nectar in absence of water, it would be an explanation for the positive results at both concentrations. Instead of diluting the food with an internal water supply, the bees in our experiment were fed with these "ideal" concentrations from the first and therefore were able to produce highest median thoracic temperatures rather constantly. Unfortunately ALTMANN and GONTARSKI published only data for two different nectar solutions. The sugar content of the crop after being fed concentrations lower than 48 % or in-between values remain unknown.

The thoracic temperatures of the worker bees without additional water supply showed a significant drop in heating at sugar concentrations lower than 45 %, and a rebound of values at 65 %. The worker bees supplied with additional water had lower thoracic temperatures in general but there was no drop in median thoracic temperatures at increasing sugar concentrations.

Two major factors seem to influence the thoracic heat production of the worker bees in our experiment: the sugar content and the water content of the food offered. If the water content is too high and the sugar content consequently low, the worker bees show low temperatures as in the present case at the two lowest nectar concentrations in both experimental groups. If the water content is too low and the sugar content consequently high, the workers show lower temperatures in the experimental group fed with nectar of 70 and 75 % sugar content without additional water. The experimental group with additional water was able to dilute the nectar as needed (Fig. 3.3 and 3.4).

The experimental group without additional water showed high temperatures at 45 % which indicates that sugar content of 45 % is enough to show steady heating performance and the water content of 55 % is not too high. Their output drops at 50 % sugar content which seems illogical, since the sugar content was high enough at 45 % already.

The oxidation of fat or sugar in energy metabolism not only requires but produces water i.e. if the honeybee thoracic muscles produce heat; they produce water as a by-product as well (BERTSCH, 1984; NICOLSON & LOUW, 1982). In flight, the water evaporates quickly and the flying bee has a lack of water in their metabolism, rather than a surplus. In heating bees in the hive, the water evaporates rather quickly by heat and air drafts produced by fanning bees. In our experiment, there was no air draft and the excess water from the thoracic heat production accumulated in the small boxes. This condition can be observed if honeybee groups that are kept in small and poorly ventilated boxes while being fed with high sugar content food. The excess water gives the worker bees a moist and "sweaty" look. (BASILE R., personal observation).

Food with 45 % sugar content seems to be a concentration which does not lead to a production of excess water. Higher sugar contents which lead to higher thoracic temperatures also lead to a higher water production. In our experimental setup this additional water did not evaporate, but most likely increased the metabolic water production of the bees and created higher humidity in the box or made them "sweat".

This negative influence of water may have been balanced out at the syrup with 65 % sugar content where the bees without additional water had their optimum (Fig. 3.1 and 3.2).

This balance between sugar and water shows that the positive correlation between sugar and heat is only true for situations were bees are either able to add water to their food or have the opportunity to get rid of excessive water from their metabolism.

The worker bees which were free to add water to their diet showed a constant increase in thoracic temperature according to the increase in sugar content of the nectar offered. The bees which had to get by with the water contained in their specific sugar concentration showed similar results for low sugar concentrations which had consequentially a relatively high water content, and for the 65 % (without additional water) and 62 and 72 % (with additional water) concentrations which were metabolized similarly well in winter bees. In addition, the workers without additional water supply reached relatively high thoracic temperatures while being fed with 45 % sugar content nectar. As aforementioned, similar concentrations were measured by ALTMANN and GONTARSKI (1963) in crop contents of bees fed with 54.1 % (± 1 %) and 80.2 % (± 1.7 %) sugar content nectar. Both concentrations might represent a target state for efficient metabolism and therefore entail high thoracic temperatures in workers without additional water. Worker bees with additional water supply showed no high thoracic temperatures at a nectar concentration of 45 % sugar content; the difference might be based on the additional water the other experimental group was supplied with or the physiological difference in the winter bee *per se*.

The opportunity to mix any amount of water needed to the offered nectar could simplify the metabolizing of high sugar content nectar and relieve the honeybees from using water bound in its tissues or hemolymph which would stress the organism.

The honeybee's water balance depends mostly on season and on ambient temperatures. Winter bees are known to regulate their water balance by the lack of a hormone from the *Corpora allata* (ALTMANN & GONTARSKI, 1963). The absence of this hormone leads to a better permeability for water in the intestines and the rectum. Winter bees recycle the "metabolic water" water which enables them to keep the feces at a lower than normal level and should make a supply of water unnecessary.

Although the effect of the hormone is associated with the ambient temperature, worker bees kept at temperatures over 20 ℃ lost "metabolic water" exponentially with rising ambient temperature (ALTMANN & GONTARSKI, 1963). By contrast, worker bees kept at temperatures lower than 20 ℃ tend to accumulate water. Our experiments were conducted at ambient temperatures of 20 ℃ ± 1, therefore the effect of the hormone should be insignificant in both our experiments.

An incalculable variable is the age of the bees used in our experiment. The summer bees in our experiments were 4 days old. This age was chosen with good cause, since younger bees are often not able to produce thoracic heat at a constantly high level and older bees tend to be under stress in small cages, which would be a behavioral difference to the winter bees which are adapted to staying in the hive over long periods of time.

The generally higher thoracic temperatures in workers without additional water temperatures compared to the workers with additional water supply might be an effect of this age related variable.

In summary, there are three major variables potentially influencing the thoracic temperatures in these experiments with different significance: the age of the bees, the season and the sugar-water ratio offered in the experimental setup.

Age might have affected individual abilities to produce heat, or their behavior in the artificial situation of being caged, resulting in stress-related or aggressive heating. We tested each nectar concentration on three different groups and found significant differences between the thoracic temperatures of almost every treatment group and the variance was low compared to the winter bees. Therefore, we assess the influence of age as relatively low in the comparison of the two setups.

The physiology and the behavior of the honeybee changes with season to a large extent (MAURIZIO, 1950; FLURI ET AL., 1987; KUNERT & CRAILSHEIM, 1988). The water balance which seems to account for the honeybees´ nutrition too, is affected by the ambient temperature and not by the season itself. Our experiments were conducted in a climate-controlled laboratory at 20 °C ±1, which is the neutral zone for the water balance influencing hormone. Therefore, we infer that season is a minor influence on water balance of the honeybees in our experiments as well.

The sugar-water ratio has a major effect on the thoracic temperature in our experiment, since concentration and thoracic temperature were correlated. There seems to be an optimal ratio of 65 % sugar and 35 % water for high thoracic heat output.

Our results show that water has a high impact on the usability of high sugar content food as it is being used in the colony to keep the brood warm or to keep the temperature at the core of the winter cluster at an adequate level. If there is enough water to dilute the sugar-rich food, the ability to produce heat increases with the sugar content. If the workers have no access to water, the heating performance is subject to fluctuations emanating from the physiological process of muscular activity itself. The metabolic water plays a significant role for thermoregulation in flying honeybees because it keeps them from overheating. In heating honeybees this by-product can cool down the heating worker and if the workers form a dense cluster and there is no constant air flow it even can remain on the body surface and cool down the heating bees.

Does sugar equal heat?

We could show that thoracic temperature and sugar content correlate positively. Since this increase was measurable in caged worker bees without an actual task, it is conceivable to assume that this correlation is not restricted to certain behaviors and therefore cannot be primarily regarded as a source of information.

3.6 Appendix – Figures and Tables

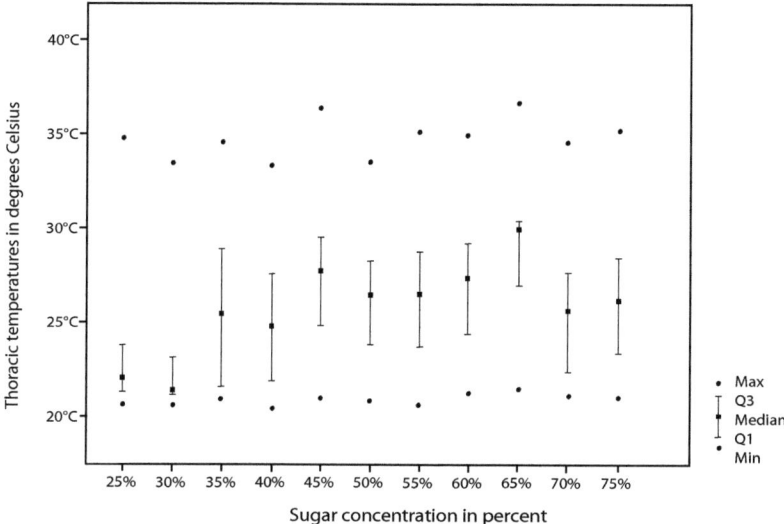

Fig 3.1 Thoracic temperatures in worker bees fed with different sugar solutions without additional water (For details and statistics see Tab. 3.1 and Suppl. Tab. 3).

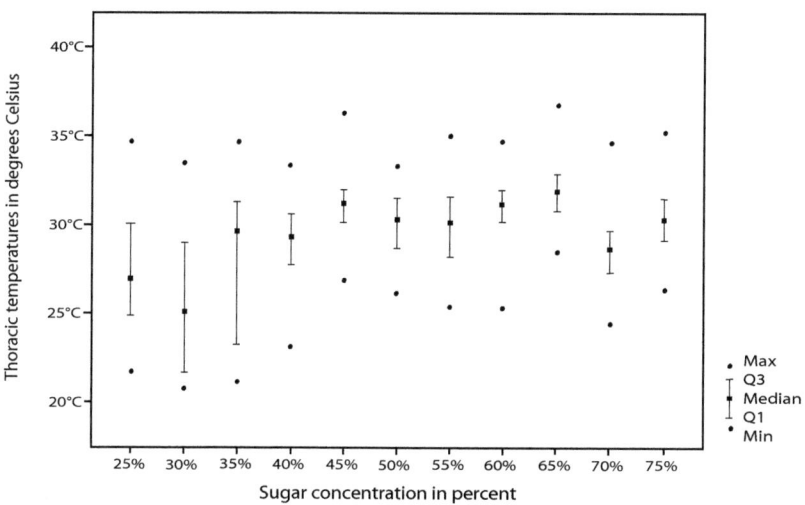

Fig 3.2 Maximal thoracic temperatures in worker bees fed with different sugar solutions without additional water (For details and statistics see Tab. 3.2 and Suppl. Tab. 4).

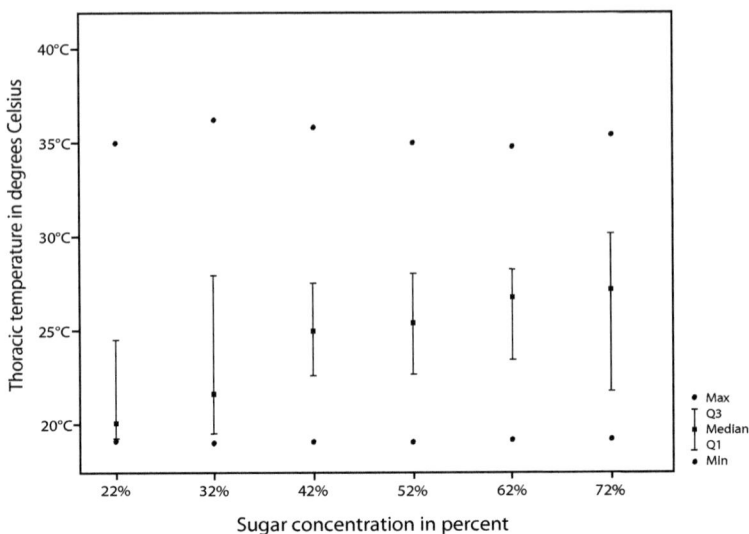

Fig 3.3 Thoracic temperatures in worker bees fed with different sugar solutions and additional water (For details and statistics see Tab. 3.3 and Suppl.Tab. 5).

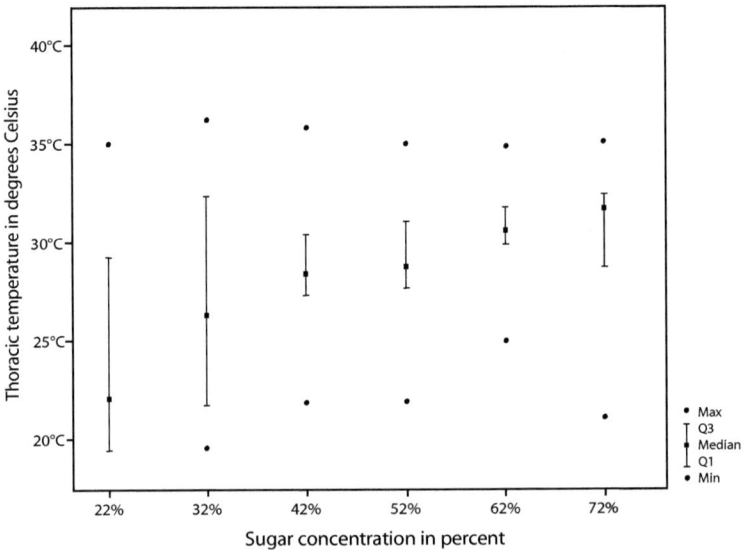

Fig 3.4 Maximum thoracic temperatures (hottest bee per group and picture) in worker bees fed with different sugar solutions and additional water (For details and statistics see Tab. 3.4 and Suppl. Tab. 6).

Does sugar equal heat?

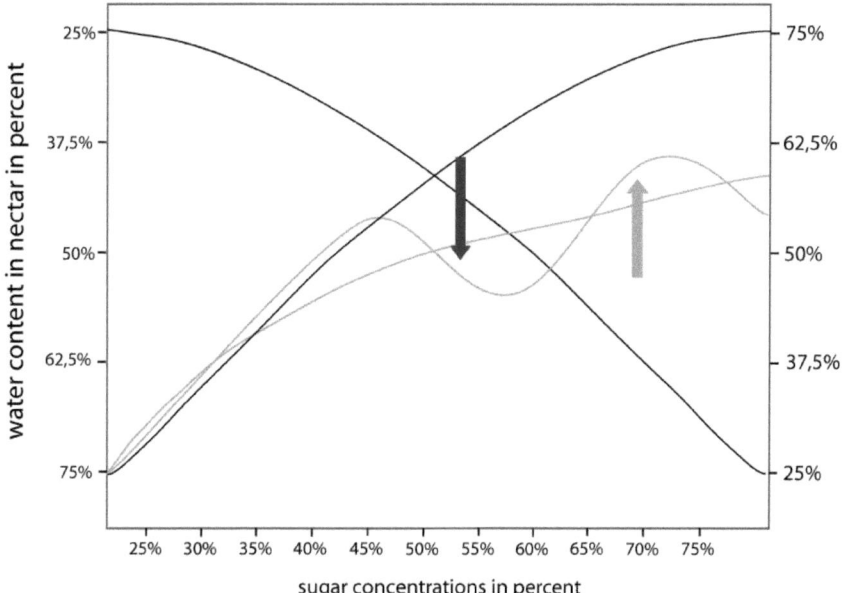

Fig 3.5 Possible influence of water, sugar and water produced by energy metabolism on the thoracic temperature of caged honeybees

black line = water content, respectively sugar content
straight grey line = thoracic heat if enough water is present
oscillating grey line = thoracic heat if no additional water is present
dark arrow = decrease in thoracic heat due to increase in metabolic water
light arrow = increase in thoracic heat due to increase in sugar

Does sugar equal heat?

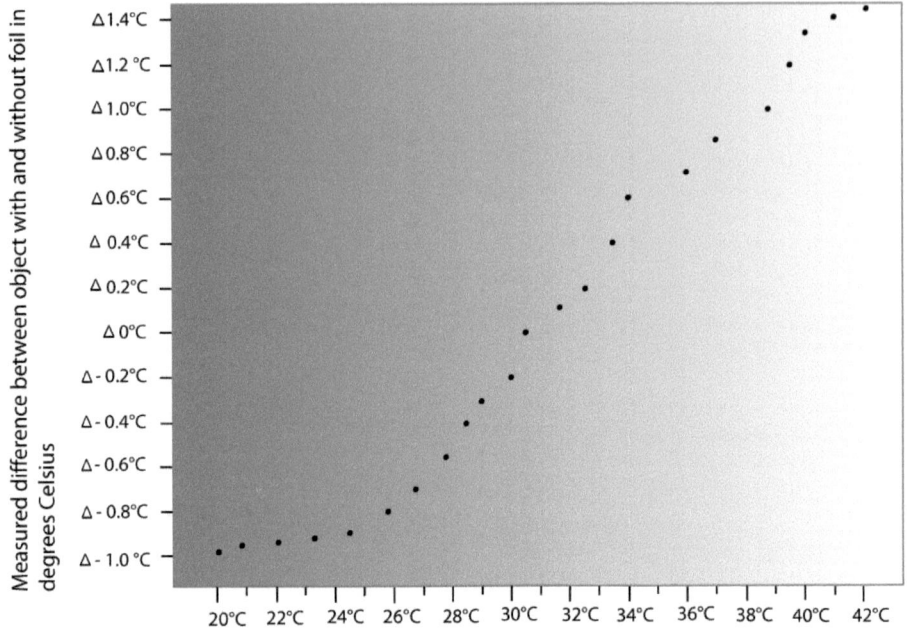

Suppl Fig. 2 Measured temperature value and the deviation produced by the foil (For details see Suppl. Tab. 2).

Sugar concentration	Median	Q1-Q3 (Δ°C)	Min	Max
25 %	22 °C	Δ 2.6 °C	20.7 °C	34.7 °C
30 %	21.5 °C	Δ 2.2 °C	20.6 °C	33.3 °C
35 %	25.5 °C	Δ 7.5 °C	20.9 °C	34.5 °C
40 %	24.7 °C	Δ 5.5 °C	20.4 °C	33.3 °C
45 %	27.8 °C	Δ 5.0 °C	20.9 °C	36.4 °C
50 %	26.7 °C	Δ 4.7 °C	20.9 °C	33.5 °C
55 %	26.7 °C	Δ 4.9 °C	20.6 °C	35.1 °C
60 %	27.2 °C	Δ 5.0 °C	21.2 °C	34.9 °C
65 %	29.9 °C	Δ 3.6 °C	21.4 °C	36.7 °C
70 %	25.7 °C	Δ 4.5 °C	21.1 °C	34.5 °C
75 %	26.2 °C	Δ 5.6 °C	21.0 °C	35.2 °C

Tab. 3.1 Thoracic temperatures (all bees per picture) of worker bee groups fed with different sugar concentrations without additional water as seen in Fig. 3.1 (For statistics see Suppl. Tab. 3).

Sugar concentration	Median	Q1-Q3 (Δ°C)	Min	Max
25 %	26.9 °C	Δ 5.0 °C	21.7 °C	34.7 °C
30 %	25.2 °C	Δ 7.4 °C	20.8 °C	33.3 °C
35 %	29.5 °C	Δ 8.1 °C	21.1 °C	34.5 °C
40 %	29.3 °C	Δ 2.8 °C	23.0 °C	33.3 °C
45 %	31.3 °C	Δ 1.9 °C	26.8 °C	36.4 °C
50 %	30.2 °C	Δ 1.9 °C	26.1 °C	33.5 °C
55 %	30.3 °C	Δ 3.3 °C	25.6 °C	35.1 °C
60 %	31.0 °C	Δ 1.7 °C	25.5 °C	34.9 °C
65 %	31.8 °C	Δ 2.0 °C	28.4 °C	36.7 °C
70 %	28.6 °C	Δ 2.2 °C	24.4 °C	34.5 °C
75 %	30.4 °C	Δ 2.2 °C	26.3 °C	35.2 °C

Tab. 3.2 Thoracic temperatures (only hottest bee per picture) of worker bee groups fed with different sugar concentrations without additional water as seen in Fig 3.2 (For statistics see Suppl. Tab. 4).

Sugar concentration	Median	Q1-Q3 (Δ°C)	Min	Max
22 %	20.2 °C	Δ 4.9 °C	19.1 °C	35.0 °C
32 %	21.7 °C	Δ 8.5 °C	19.0 °C	36.2 °C
42 %	25.0 °C	Δ 5.0 °C	19.0 °C	35.8 °C
52 %	25.4 °C	Δ 5.4 °C	19.0 °C	35.0 °C
62 %	26.8 °C	Δ 5.7 °C	19.1 °C	34.8 °C
72 %	27.1 °C	Δ 8.4 °C	19.1 °C	35.4 °C

Tab. 3.3 Thoracic temperatures (only hottest bee per picture) of worker bee groups fed with different sugar concentrations and additional water as seen in Fig. 3.3 (For statistics see Suppl. Tab. 5).

Sugar concentration	Median	Q1-Q3 (Δ°C)	Min	Max
22 %	22.0 °C	Δ 9.5 °C	19.4 °C	35.0 °C
32 %	26.3 °C	Δ 10.7 °C	19.5 °C	36.2 °C
42 %	28.3 °C	Δ 3.3 °C	21.9 °C	35.8 °C
52 %	28.6 °C	Δ 5.4 °C	21.9 °C	35.0 °C
62 %	30.7 °C	Δ 2.1 °C	25.0 °C	34.8 °C
72 %	31.9 °C	Δ 4.3 °C	21.1 °C	35.4 °C

Tab. 3.4 Thoracic temperatures (only hottest bee per picture) of worker bee groups fed with different sugar concentrations and additional water as seen in Fig. 3.4 (For statistics see Suppl. Tab. 6).

Measured Temp (with foil)	Difference (without foil)
20.0 ℃ – 24.4 ℃	Δ - 0.9 ℃
24.5 ℃ – 25.7 ℃	Δ - 0.8 ℃
25.8 ℃ – 26.9 ℃	Δ - 0.7 ℃
27.0 ℃ – 27.5 ℃	Δ - 0.6 ℃
27.6 ℃ – 28.5 ℃	Δ - 0.4 ℃
28.6 ℃ – 29.0 ℃	Δ - 0.3 ℃
29.1 ℃ – 30.0 ℃	Δ - 0.2 ℃
30.1 ℃ – 31.0 ℃	Δ +/- 0 ℃
31.1 ℃ – 32.6 ℃	Δ + 0.2 ℃
32.7 ℃ – 33.5 ℃	Δ + 0.4 ℃
33.6 ℃ – 34.1 ℃	Δ + 0.6 ℃
34.2 ℃ – 36.0 ℃	Δ + 0.7 ℃
36.1 ℃ – 36.9 ℃	Δ + 0.8 ℃
37.0 ℃ – 38.7 ℃	Δ + 1.0 ℃
38.8 ℃ – 39.5 ℃	Δ + 1.2 ℃
39,6 ℃ – 42.0 ℃	Δ + 1.3 ℃

Suppl. Tab. 2 Differences between measured temperature with and without foil, influence (Δ ℃) produced by the foil (error) which has to be added or subtracted from the measured temperature.

Does sugar equal heat?

Sugar concentration in %	Sugar concentration in mol
22 %	0.63 mol
25 %	0.71 mol
30 %	0.86 mol
32 %	0.91 mol
35 %	1.0 mol
40 %	1.14 mol
42 %	1.2 mol
45 %	1.28 mol
50 %	1.43 mol
52 %	1.48 mol
55 %	1.57 mol
60 %	1.71 mol
62 %	1.77 mol
65 %	1.86 mol
70 %	2.0 mol
72 %	2.06 mol
75 %	2.14 mol

Suppl. Tab. 3 Sugar concentrations used in chapter 3 in percent and mol

Does sugar equal heat?

Sugar concentration in %	30 %	35 %	40 %	45 %	50 %	55 %	60 %	65 %	70 %	75 %
25 %	***	***	***	***	***	***	***	***	***	***
30 %	-	***	***	***	***	***	***	***	***	***
35 %	-	-	*	***	***	***	***	***	n.s.	***
40 %	-	-	-	***	***	***	***	***	***	***
45 %	-	-	-	-	***	***	***	***	***	***
50 %	-	-	-	-	-	n.s.	***	***	***	**
55 %	-	-	-	-	-	-	***	***	***	***
60 %	-	-	-	-	-	-	-	***	***	***
65 %	-	-	-	-	-	-	-	-	***	***
70 %	-	-	-	-	-	-	-	-	-	**

Suppl. Tab. 4 Multiple comparisons of all groups: p-levels for groups of worker bees without additional water.

(Kruskal-Wallis Anova: $n_{overall}$=40.310; $n_{25\%}$=4020; $n_{30\%}$=3600; $n_{35\%}$=3600; $n_{40\%}$=3410; $n_{45\%}$=3600; $n_{50\%}$=3600; $n_{55\%}$=4210; $n_{60\%}$=3600; $n_{65\%}$=3470; $n_{70\%}$=3600; $n_{75\%}$=3600; H=9320, 242; p<0.0001)

Does sugar equal heat?

Sugar concentration in %	30 %	35 %	40 %	45 %	50 %	55 %	60 %	65 %	70 %	75 %
25 %	**	***	***	***	***	***	***	***	n.s.	***
30 %	-	***	***	***	***	***	***	***	***	***
35 %	-	-	n.s.	***	**	***	***	***	n.s.	***
40 %	-	-	-	***	*	***	***	***	*	***
45 %	-	-	-	-	***	***	n.s.	***	***	**
50 %	-	-	-	-	-	n.s.	***	***	***	n.s.
55 %	-	-	-	-	-	-	***	***	***	n.s.
60 %	-	-	-	-	-	-	-	***	***	**
65 %	-	-	-	-	-	-	-	-	***	***
70 %	-	-	-	-	-	-	-	-	-	***

Suppl. Tab. 5 Multiple comparisons for all groups: p-levels for groups of worker bees without additional water (only hottest bee per group picture).

(Kruskal-Wallis Anova: $n_{overall}=4031$; $n_{25\%}=402$; $n_{30\%}=360$; $n_{35\%}=360$; $n_{40\%}=341$; $n_{45\%}=360$; $n_{50\%}=360$; $n_{55\%}=421$; $n_{60\%}=360$; $n_{65\%}=347$; $n_{70\%}=360$; $n_{75\%}=360$; H=1173, 559; p<0.0001)

Sugar concentration in %	32 %	42 %	52 %	62 %	72 %
22 %	***	***	***	***	***
32 %	-	***	***	***	***
42 %	-	-	**	***	***
52 %	-	-	-	***	***
62 %	-	-	-	-	n.s.

Suppl. Tab. 6 Multiple comparisons for all groups: p-levels for groups of worker bees with additional water

(Kruskal-Wallis Anova: $n_{overall}$= 21780; each group n=3630; H = 3033,300; p<0.0001)

Sugar concentration in %	32 %	42 %	52 %	62 %	72 %
22 %	***	***	***	***	***
32 %	-	n.s.	n.s.	***	***
42 %	-	-	n.s.	***	***
52 %	-	-	-	***	***
62 %	-	-	-	-	n.s.

Suppl. Tab. 7 Multiple comparisons for all groups: p-levels for groups of worker bees with additional water (only hottest bee per group and picture)

(Kruskal-Wallis Anova: $n_{overall}$= 2178; each group n=363; H = 401,6742; p<0.0001)

4. Trophallactic activities in the brood nest - Heaters Get Supplied with High Performance Fuel

4.1 Abstract

Honeybees actively regulate their brood temperature by heating to keep the temperature between 33 °C to 36 °C if ambient temperatures are lower. Heat is generated by vibrating the flight muscles and is physiologically approximate to flying. Heating rapidly depletes the worker's internal energy; therefore heating performance is limited by the honey that is ingested before the heating process. Stored honey is the predefined fuel for flying and heating, but it is stored at a distance from the brood comb, causing a potential logistic problem of efficient energy supply in the brood area.

Our study focused on the behavior and the thoracic temperature of the participants in trophallactic food exchanges on the brood comb. The brood area is the center of heating activity in the hive, and therefore the region of highest energy demand. We found that 85.5 % of the recipients in a trophallactic food exchange have a higher thoracic temperature during feeding contacts than donors and after the feeding contact the former engage in brood heating more often. The donor bees have lower thoracic temperature and shuttle constantly between honey stores and the brood comb where they transfer the stored honey to heating bees.

Providing heat-emitting workers with small doses of high performance fuel contributes to an economic distribution of resources consistent with physiological conditions of the bees and the ecological requirements of the hive. Only if the basic need for keeping up their own energy metabolism is guaranteed, individuals can engage in other tasks which are not directly related to keeping up their own energy metabolism, e.g. heating. The trophallaxis based system of output-related supply is essential to provide the energy-intensive brood warming activity. The emerging independence from ambient temperatures is not only beneficial for brood rearing during times of sudden cold spells, but also enables the honeybees in temperate regions to raise brood in early spring and might be the decisive factor for the occurrence of honeybees in temperate climates in general.

4.2 Introduction

Worker bees of *Apis mellifera* maintain the temperature of their pupae between 33 °C and 36 °C by heating or cooling (HIMMER, 1927; SEELEY & HEINRICH, 1981; ESCH & GOLLER, 1991; HEINRICH, 1993). If temperatures are not kept within these limits, the results may be brain damage and losses in behavioral capability (TAUTZ ET AL., 2003; GROH ET AL., 2004). The heating bees station themselves on the brood comb where they transfer heat either by pressing their hot thoraces onto capped cells (BUJOK ET AL., 2002), or by crawling head first into empty cells within the brood comb to heat neighbouring brood from the side (KLEINHENZ ET AL., 2003). This uninterrupted cell-heating activity was observed to last up to 32.9min by KLEINHENZ ET AL. (2003). Heat production in honeybees is done by "shivering" the flight muscles. During this shivering, wing and thoracic vibrations are generally not detectable and the bees may appear to be quiet and "at rest". The thoracic heat is a by-product of flight during which up to 60 % of the energy is released as heat or as JOSEPHSON (1981) put it: *"It [Insect flight] efficiently converts chemical energy to mechanical power and, because of biochemical inefficiencies, heat."*

Brood heating and flying rapidly consumes the energy reserves of a worker bee. This can be concluded from the equality in oxygen consumption by the bees for both activities, which is 1.16µl/g/min during flight muscle shivering and 1.14µl/g/min during flight (HEINRICH, 1993).

Since honeybees use mostly sugar as energy substrate for muscular activity (JONGBLOED & WIERSMA, 1934; LOH & HERAN, 1970; SACKTOR, 1970; ROTHE & NACHTIGALL, 1989), the level of glycogen in the hemolymph must be kept high to provide an adequate fuel supply for the heat-generating flight muscles (CRAILSHEIM, 1988) which are the most metabolically active tissues known (SOUTHWICK & HELDMAIER, 1987).

In honeybees, food is stored in the crop or "honey stomach". Any liquid from the crop (nectar, water, honey) can be regurgitated and deposited in cells or transferred to other bees. The crop has a sphincter muscle, the ventriculus, which works as a valve that can release food doses into the midgut, where it is transferred into the bloodstream (BLATT & ROCES, 2001). A crop load of sugar solution can provide a bee with food for several hours. Nevertheless, even inactive, caged honeybees with a full crop held at room temperature die within 7h after being separated from their food source (HEINRICH, 1993). A physiologically challenging activity like flying or heating will consume their sugar fuel even faster, so the crop content and its sugar concentration consequently reflects the demand of the upcoming task (NIXON & RIBBANDS, 1952; CRAILSHEIM, 1988).

The brood area where the heating activity takes place is usually situated in the center of the comb and roughly surrounded by pollen containing cells. The stored honey, which is the best source of carbohydrates in the hive, is kept at the upper corners of the comb which is extended towards the center in the course of time (SEELEY & MORSE, 1976). The brood comb and the honeycomb are separated from each other by several empty cells and the pollen circle. These empty cells between brood and food are kept empty as long as there is enough space for additional nectar deposits on the honeycomb and additional brood cells on the brood comb (Fig. 4.1). SEELEY (1982) suspected that the spatial segregation of brood nest and food storage regions was initially advantageous because it facilitates brood incubation and probably helped to economise in nest construction.

The separation of brood nest and food storage creates a spatiotemporal gap between brood and food which must be bridged by the heating bees for a regular reload of honey, because of their increased energy requirement. The expense of heat loss for a heating bee, which leaves the brood nest, is irrespective of the distance it needs to bridge between the brood nest and the honeycomb. Since the clustering on the brood comb serves the same purpose like in winter cluster, to reduce and prevent heat loss, every movement of heating bees in or out reduces the insulation efficiency.

Indeed, SOUTHWICK and HELDMAIER (1987) wrote that the efficiency of tight clustering in winter can reduce the effective area of heat exchange by as much as 88 %. Brood incubation in that manner economises the active heating activity of the bees by means of a more economic organisation of the cluster.

The food intake to fuel a honeybee's activity is either done by the worker bee itself, i.e. taking up food from the honeycomb, or by getting fed by another bee which regurgitates food from her crop and transfers it mouth to mouth. This feeding activity between two individuals is called trophallaxis (FREE, 1956).

There are two ways of how a trophallactic contact can start: Firstly, a bee can beg for food by extending its proboscis and thrusting its tip towards the mouthparts of another bee. If the begging bee is successful, the other bee responds by regurgitating food and thereby is initiating a trophallactic contact. Secondly, a bee can offer food by opening its mandibles, raising the proximal part of its proboscis and regurgitate a droplet of food that is displayed between the mandibles and the proboscis (FREE, 1956). If a recipient bee touches that droplet with its antennae and then thrusts its proboscis between the mouthparts of the donor, this results as well in a trophallactic contact (MONTAGNER & PAIN, 1971).

Not all trophallactic contacts in the hive are feeding contacts. Nectar foragers for example, make trophallactic contacts with nectar receiving bees. These contacts are used more for nectar transport rather than for nourishment of the recipient, as the passed on nectar is finally deposited in the nectar cells and usually not consumed by the nectar receiving bee.

Trophallactic contacts occur all over the hive, but are carried out more frequently on the brood comb (SEELEY, 1982). These feeding contacts on the brood comb are intended for nourishment of the recipient with the utmost likelihood. The nectar passing contacts between foragers and nectar receiving bees are usually limited to an area near the hive entrance (SEELEY, 1982) which is distant from the brood comb where the observations in this study were focused (Fig. 4.1).

The task a worker is about to fulfil and the energetic requirements for the different activities in honeybees are unequal (NIXON & RIBBANDS, 1952; CRAILSHEIM, 1988). This means that foraging or heating bees spend more energy than for example pollen storing or cell cleaning bees. Consequently, the task partitioning system requires a sort of resource management that assures an ideal distribution of the available stocks to the worker bees that are performing the more strenuous activities.

A honeybee that is preparing to go on a foraging flight needs a certain amount of food to reach her destination and return to the hive. Foragers do not fill their crops at the honeycomb or use the nectar they collect for the flight. BRANDSTETTER ET AL. (1988) found that foragers get refuelled via trophallaxis between foraging flights by worker bees in the hive.

In the same way that the foragers have to balance the distance and fly against the fuel needs (BEUTLER, 1950; SACKTOR, 1970), heater bees need a mechanism to bridge the spatiotemporal gap between heating the brood and replenishing their energy resources. Like in foragers, the area where their task has to take place is separated from the honeycomb where the energy they need to fulfil their task is stored.

Given that the distribution of laboriously collected energy resources has to be in line with demand as it is in foraging flights (SCHOLZE ET AL., 1964; CRAILSHEIM, 1988) raises the question how the heating performance and the nourishment tasks are regulated in an efficient way without misspending stored honey or wasting produced heat by leaving the heating cluster. Therefore, the trophallactic food dispersal and heating have to be interrelated by a task sharing system, and must be distinguishable by the behavior and the temperature of the participants in a trophallactic contact.

4.3 Materials and Methods

All observations were made at the Beestation of Würzburg University (Biocenter) from May to July 2005 with *Apis mellifera carnica* in four standard two-frame observation hives (3000 – 4000 bees) in a shaded room under red light conditions. Unrelated artificially inseminated queens (10 to 12 drones) were heading these observation hives.

4.3.1. Behavioral observations

Tracking of the honeybees was done at two-frame observation hives through a transparent foil placed over the pane and a stop watch. The participants of a trophallactic contact were identified as donors or receivers and tracked for as long as possible but not exceeding 20min. The duration of 20min was used for the observation since the probability that the behavior after the trophallactic contact and the role the participant had are connected as cause and effect, decreases over time. We decided that an action that takes place more than 20min after the feeding contact can no longer be logically connected as cause with the preceded activity of feeding or getting fed as effect.

These tracks were sketched with a marker on the foil. The different behaviors and their duration were listed on the foil with graphic symbols and the duration of the performed behaviors noted. The tracking, together with a detailed map of the hive (i.e. cell contents, respectively brood comb, honeycomb, empty cells etc.) give full information about time, space and behavior of the observed bees at all times (Fig 4.1).

The behavioral patterns that are associated with food transmission (RIBBANDS, 1953; CRAILSHEIM, 1988) and heat production (BUJOK ET AL., 2002) were distinguished and the duration was recorded and noted as percentages of the whole observation time. The observed behavior patterns include trophallaxis which was differentiated into donating food and receiving food as described by FREE (1956); sticking the head into a cell containing honey which is necessary to unload or gather honey, and cell-heating i.e. crawling into an empty cell while continuously pumping abdomen as described by KLEINHENZ ET AL. (2003). In addition, the absolute percentage of behavioral actions of donors and recipients, that is whether an observed bee ever showed a special behavior or not, was noted.

We counted exclusively cell-heating behavior as "heating", because it provides a definite indication of heating behavior without measuring the thoracic temperature of the bee and because the second experiment with the thermal imaging camera provides accurate data concerning this issue.

4.3.2 Thermal imaging

In addition to the behavioral data, thermal images of trophallactic contacts on the brood comb were recorded. Both experiments were performed at the same hives and the same period (from May to July 2005). With the thermal imaging camera S40 from FLIR Systems ^{TM}, close up shots were taken of the capped brood area at two-frame observation hives in a darkened room. The panes were replaced by heat radiation permeable foil. We used the plastic wrap "Cling Wrap"© made of 100% polyethylene manufactured by the U.S. American company GLAD™. Since the foil still alters the radiation averting from the object and the radiation measured by the camera, we needed to define the error produced at different temperatures. (See also chapter 3, Suppl. Tab. 3 and Suppl. Fig. 2) The error produced by the foil is non-linear, therefore every thoracic temperature measured had to be specifically corrected (see Suppl. Fig. 2 and Suppl. Tab. 2).

The film footage consists of 32h (128 sequences à 15min) taken on 15 different days between 11:00a.m. and 14:00p.m. After each sequence, the camera was moved to another part of the brood comb in order to avoid pseudo-replicates. The recorded images were analysed with the software ThermaCam^{TM} Researcher Pro 2.7 from FLIR Systems^{TM}. We used an emissivity of 0.97 of the honeybee cuticle as it is described by STABENTHEINER and SCHMARANZER (1987).

The collected data includes the thoracic temperature of the participants at the beginning of the contact and the duration of the food exchange. The temperature was only measured at the beginning of the contact, as our study focuses on the thermal cause and not on the effect of a trophallactic contact. Using thermal imaging, the thoracic temperature of donor and recipient worker bees in the capped brood area were measured at the initiation of the food transfer, as was the duration of the contact and the temperature difference between a pair.

Donor and recipient trophallactic participants were recognised by the way in which they manipulate their mouthparts (FREE, 1956) and were categorised by the direction of the food exchange.

The thorax has a three-dimensional shape which produces differences in thermal radiation at the edges. We measured the maximal thoracic temperature which is usually the middle of the thoracic surface in top view (Fig. 4.2A).

Behavioral patterns like feeding and getting fed can only be distinguished in close up shots that produce a frame of 7.5x11cm. The bees often leave the frame during observation, so only the behavior of almost stationary bees would be observable.

Worker bees (especially donors) tend to leave the frame of the camera, when taking close up shots, which are necessary to distinguish behavioral patterns; therefore the tracking had to be done by an experimenter and not with the thermovision camera.

All statistical analyses were performed using the statistical package Statistica 8©.

4.4 Results

4.4.1 Behavior of donors and recipients

Individual tracking began with a trophallactic couple in the capped brood area, where the participants in the feeding contact could be clearly defined by their individual behavior patterns as donors and recipients and therefore categorised in one of the two groups. In order to establish that the donors were indeed fetching food from the honeycomb and that the recipients were heating the brood comb, we observed the bees up to 20min or until we lost track of them.

Both donors and recipients were observed to cover the distance between brood comb and honeycomb; however the frequency of this activity was significantly higher in donors than it was in recipients after a trophallactic contact (Fig. 4.3). Donors tend to move back and forth between brood comb and honeycomb twice on average in one observation, while recipients usually stay on the brood comb (Fig. 4.1 and Fig. 4.3).

Trophallactic activities in the brood nest

A set of behavioral patterns was performed more often by workers that were acting as donors in the first observed contact. The repeated donation of food (median in donors 8.9 % of the observation time, median in recipients 0 % of the observation time) and the insertion of the head into a honey cell that is required to refill the crop with honey (median in donors 5.3 % of the observation time, median in recipients 0 % of the observation time) were almost exclusively performed by donors (Fig. 4.4A, C and Fig. 4.5A, C).

To exclude the possibility that the observed bees were actually nurse bees, and that therefore the feeding of the other bees actually is part of nursing behavior, we compared that nursing behavior for donors and recipient bees. There were no significant differences in the nursing behavior. The median in donors was 1.9 % (quartiles: Q1=0 %, Q3=6.8 %) of the observation time; the median in recipients was 0 % (quartiles: Q1=0 %, Q3=4.2 %) of the observation time. This indicates that the trophallactic interactions represent a different distinct behavioral pattern (Mann-Whitney-U-Test: Donors (n=75), Recipients (n=75) inserting head into larva containing cell: U=2339, Z=1.78, p<0.07).

Workers that were acting as recipients in the first observed contact exhibited the repeated reception of food (median in recipients=9.5 % of the observation time, median in donors=0 % of the observation time), and cell-heating (median in recipients=7.2 % of the observation time; median in donors=0 % of the observation time) about a longer period of time and more frequently (Fig. 4.4B, D and Fig. 4.5B, D).

Some bees „switched" tasks during the observation, i.e. there were bees acting as donors in the first observed feeding contact, but were observed getting fed by other bees (17.3 % / 13 bees out of 75) or engaging in cell-heating afterwards (16 % /12 bees out of 75). Likewise, there were recipients which engaged in feeding other workers (30.6 % 23 bees out of 75), or gathered honey after the first observed contact (18.6 % / 13 bees out of 75) (Fig. 4.2B). The absolute percentage of behaviors and the percentage of behaviors shown by donors or recipients set of against the observation time can be seen in Figure 4.5. Nevertheless, the median percentage of the time spent with the "wrong" task in donors or recipients were still 0 %.

Additionally, the correlation between the different behaviors in all observed bees, donors and recipients, were calculated.

Gathering honey and donating food significantly positively correlate, meaning that the more time food donations were registered in an individual, the more time the same individual spent with the honey gathering activity (Spearman coefficient: n=150, R=0.49, p<0.05).

Gathering honey and receiving food significantly negatively correlate, meaning that the more time food receptions were registered in an individual, the less time the same individual spent with gathering honey (Spearman coefficient: n=150, R=-0.30, p<0.05).

Cell-heating and food reception significantly positively correlates, meaning that the more time food receiving was registered in an individual, the more time the same individual spent with cell-heating activity (Spearman coefficient: n=150, R=0.37, p<0.05).

Cell-heating and food donation significantly negatively correlate, meaning that the less time the same individual spent with cell-heating activity, the more time the individual spent donating food (Spearman coefficient: n=150, R=-0.49, p<0.05).

4.4.2 Thoracic temperature and trophallaxis

In 85.5 % (2310 in numbers) of all 2700 observed feeding contacts, the recipients exhibited a higher thoracic temperature than the donors (X^2-test: X=2730.67, p<0.001). The thoracic temperatures of the recipients (median temperature: 35.6 °C) were significantly higher than that of the donors (median temperature: 34.6 °C) at the initiation of a trophallactic contact (Fig. 4.2B); meaning that generally hotter bees received food from cooler donors.

The duration of contacts varied from a few seconds to over three minutes. Most contacts (55 %) lasted three seconds or less. The longer the duration of the interaction, the less often such interaction was observed (Fig. 4.6).

The frequent feeding activity on the brood comb supplied this study with a high number of analysed feeding contacts (n=2700). The power analysis shows that the result of the ascertained temperature difference in the trophallactic participants can be regarded as highly reliable (Power Calculation: Alpha 0.05, Power 1.0).

4.5 Discussion

The individual tracking of trophallactic participants showed significant differences in the behavior of donors and recipients:

Donors spend significantly more time in feeding other bees and sticking their heads into honey cells than recipients did (Fig. 4.4 and Fig. 4.5) Recipients significantly more often performed cell heating tasks.

The positive correlation between gathering honey and donating food in all observed bees supports these findings. Therefore, our results show that trophallactic food dispersal and brood heating are linked tasks, which are visibly distinct from each other. The behavior and temperature of the participants in a trophallactic contact are clearly distinguishable and characteristic for the performed tasks.

Gathering honey in one part of the hive and redistributing it elsewhere cannot be considered as random, especially since donor bees cover the distance bidirectionally, i.e. they donate food in the brood area, take up food on the honeycomb, then return to the brood area and donate food again, while recipients stay on the brood comb (Fig. 4.1 and Fig. 4.3). Some donors were observed to shuttle six times between brood and food within the 20min of observation. The speculation that donor bees are nurse bees who feed workers as a by-product on their way to their "real" task of larvae feeding can be regarded as unfounded. The insertion of the head into a larva containing cell which is the basis of carrying out the larvae feeding task, was not performed more often or for a longer period of time by donors as it was performed by the recipients. In addition, the observation of the trophallactic contacts which were used to distinguish between donors and recipients in this study was restricted to the capped brood area and not to the open brood area, where nurse bees usually are active (RÖSCH, 1925; LINDAUER, 1952).

Recipients spend significantly more time getting fed and heating in cells (Fig. 4.4 and Fig. 4.5). The positive correlation between cell-heating and receiving food supports these findings. Therefore, food ingestion by trophallaxis in the brood area is significantly coupled to cell-heating behavior. The difference in thoracic temperatures of donors and recipients confirms this interrelation between feeding and heating. Brief feeding contacts often lead to discussion as to whether they can be counted as real trophallactic transmission or not (FARINA & WAINSELBOIM, 2001A). In this work, the feeding act was only counted as trophallaxis if the direction of the transmission could be detected by the way the bees manipulated their mouthparts (FREE, 1956) and if there was a detectable food transmission. FARINA and WAINSELBOIM (2005) used a similar technique to demonstrate that there are food transmissions even during brief trophallactic contacts.

The thermal imaging data supports the behavioral data by giving information about the thoracic temperatures of the participants in the thermal images at the starting point of the behavioral observation.

The food intake before and in between heat production fuels the heating activity of the recipients. Especially the in-cell-heating performance that follows is limited by the sugar fuel the bee was loaded with shortly before for two reasons:

Firstly, there cannot be any trophallactic contacts with a bee that has climbed head first into an empty cell and usually there is no food deposited in these empty cells. Secondly, only very small amounts of glycogen can be stored in the flight muscle (NEUKIRCH, 1982; PANZENBÖCK & CRAILSHEIM, 1997). Heat production in the honeybee is subject to strict physiological conditions: the glycogen that is required for the metabolic efficiency of the flight muscle can be transferred only by passing the ventriculus and enters the midgut, i.e. sugar that is stored in the crop cannot enter the bloodstream directly (CRAILSHEIM, 1988). Honeybees depend almost exclusively on intestinal and hemolymph energy supplies for energetically demanding activities like heating or flying, since fat or protein can only be metabolized to a very small amount in order to increase the blood sugar level unlike in vertebrates (JOHN, 1958; PANZENBÖCK & CRAILSHEIM, 1997; MICHEU ET AL., 2000; BLATT & ROCES, 2001).

Younger bees which are usually performing the in-hive tasks including brood care (Rösch, 1925; Lindauer, 1952), have an extremely high level of glycogen, which can only be stored in limited amounts in the flight muscle (NEUKIRCH, 1982; PANZENBÖCK & CRAILSHEIM, 1997).

Therefore, the need to refuel in heating bees follows from the activity that requires a lot of energy and the limited opportunities of increasing the blood sugar level in honeybees. The physiological limits of the flight muscle, its low energy capacity and high metabolism, could cause the honeybees´ consistent behavior at either donating or receiving food in small doses and quick succession as we described in our behavioral data.

Our findings that heating bees on the brood comb are fed with honey correspond with NIXON and RIBBANDS (1952) data which showed that honeybees in the brood area are least often fed with freshly collected nectar. The radioactive nectar the foragers collected in their study was distributed all over the hive to a certain extent, but the bees on the brood comb had the lowest radioactive load, which leads to the conclusion that if they are not fed with freshly collected nectar, it must be stored honey they are supplied with. Assuming that stored honey which is fed to the bees on the brood comb has higher sugar content than freshly collected nectar, the short contacts the donor bees perform with many different bees on the brood comb might be intended to provide many heat producing bees with small doses of high performance fuel. The fact that most trophallactic contacts on the capped brood were brief and the couples on the capped brood consisted of a donor with a lower thoracic temperature and a recipient with a higher thoracic temperature speaks well for this conclusion (Fig. 4.2).

The ingestion of food which is high in sugar content might even have a direct influence on the thoracic temperature of the recipients. The connection between sugar content and thoracic temperature was described by STABENTHEINER and SCHMARANZER (1987). They found a positive correlation between sugar content of food with an increase of thoracic temperature for ingestion at the feeding place outside the hive. STABENTHEINER (1996) added the confirmation that the increase in thoracic temperature depending on the food source correlated with dancing, walking and trophallaxis, but he did not specify the area in the hive where the feeding contacts took place nor did he differentiate between donating and receiving a food transmission.

A contrary correlation for behavior and thoracic temperature was described by FARINA and WAINSELBOIM (2001B). They described a higher thoracic temperature in returning nectar foragers and a cooler thoracic temperature in nectar receiver bees at the beginning of a trophallactic contact (n=69). Our observation, on the other hand, showed a contrary effect in thoracic temperature concerning donors and recipients in trophallactic contacts on the brood comb in 2390 of 2700 cases (Fig. 4.2A, B). This is not surprising as the higher thoracic temperature at the beginning of the trophallactic contact measured by FARINA and WAINSELBOIM (2001B) is probably a side effect of the flying action the returning nectar forager has carried out shortly before. As aforementioned, our observations with the thermal imaging camera were strictly related to the brood comb, which is not identical to the hive entrance where the nectar transfer between returning foragers and nectar receiving bees usually takes place (SEELEY, 1989). The donors in our behavioral observation shuttled between honeycomb and brood comb, and are therefore not very likely to be returning nectar foragers.

Cooler bees that are occupied with supplying and regurgitation of food with high sugar content and hotter bees that are occupied with cell-heating show clear task sharing, which is beneficial for the whole colony. Donor bees that distribute small doses of high performance fuel to many heat emitting bees, instead of supplying bees randomly on their way across the hive, increases the efficiency of the feeding task and of the heating bees.

Why heating bees prefer to stay and continue performing the energy consuming task uninterruptedly, which is very likely to reduce their life span (NEUKIRCH, 1982) is unknown.

It is an established fact that the continuous heating of the recipient bees entails an increase in brood rearing efficiency by bringing nourishment to the active heaters whose energy capacity is low and so avoiding down time by leaving the brood to reload on honey. In addition, the reduction of movement on the brood comb improves the efficiency of insulation by keeping the cluster together which is necessary for brood comb heating and insulation against heat loss (SACKTOR, 1970; KRONENBERG & HELLER, 1982; SOUTHWICK & HELDMAIER, 1987).

This task partitioning system which provides heat-producing workers with small doses of high performance fuel, contributes to a highly economical resource management that is in line with physiological conditions of the bees and the ecological requirements of the hive. Moreover, unlike bumblebees were trophallaxis between workers is lacking (DORNHAUS ET AL., 1998), this trophallactic behavior is the underlying mechanism to develop a sophisticated task partitioning. The resulting economical resource management might be one of the factors favouring the evolution of perennial bee colonies in temperate regions.

4.6 Appendix – Figures and Tables

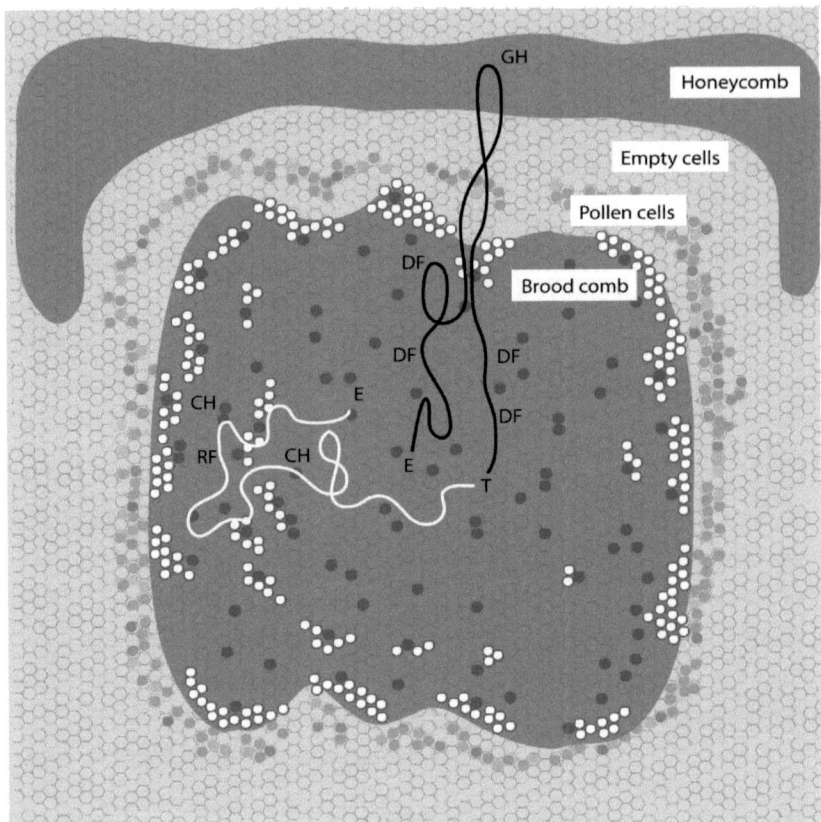

Fig. 4.1 Map depicting the position of brood comb and honeycomb in a two-frame observation hive.

Different areas and the cell contents in the hive with tracks of a donor bee (black line) and a receiving bee (white line) after the trophallactic contact give an example of how the data was collected. The "shuttling" behavior of the donor bee is noticeable by the tracks which lead to the honeycomb and back to the brood comb which is how they bridge the spatiotemporal gap between brood and food while recipients stay on the broodcomb.
Cell contents in the brood area are differentiated by colour: dark grey=empty cell in the brood area, white=larvae or egg-containing cell, lighter grey=capped brood cell.

The different behaviors of donor and receiving bee are abbreviated:
T=trophallaxis (beginning of observation), RF=receiving food, DF=donating food, CH=cell-heating, GH=gathering honey, E=end of observation.

Trophallactic activities in the brood nest

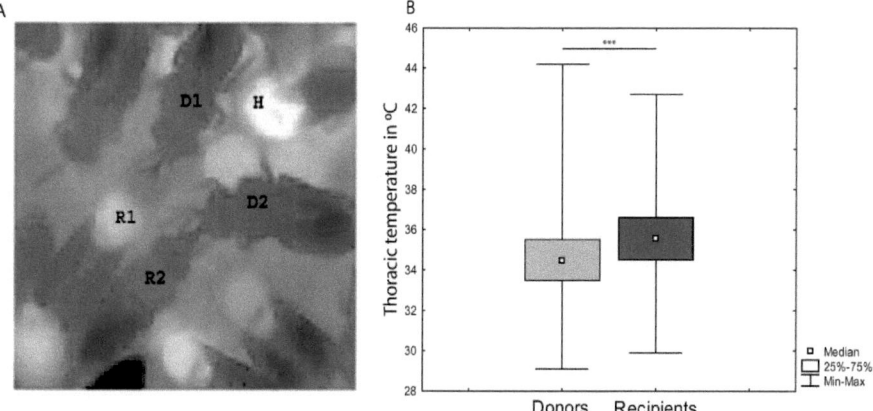

Fig. 4.2 Coherence between thoracic temperature and role in trophallactic contact

A Fixed-image detail from thermal imaging camera of trophallactic contacts

Display detail of feeding contact in the capped brood area: recipients R1 (T=37.5 °C) and R2 (T=35.8 °C) are getting fed by donors D1 (T=33.8 °C) and D2 (T=34.4 °C) while a cell-heating bee H (T=40.8 °C) climbs into a cell after a food transfer.
The direction of the transmission can be recognized by the way in which the bees manipulate their mouthparts and by the changing color of the proboscis. The temperature of the transmitted food leads to a change in color in the thermal image. In this case the transmitted food is cooler than the recipient's mouthparts and they change color from lighter grey to darker grey (white = hot, black = cold).

B Differences of thoracic temperature in donors and recipient on the brood comb

(Mann-Whitney-U-Test: Donors (n=1350), (median temperature: 34.6 °C, quartiles: Q1=33.5 °C, Q3=35.5 °C, Min=29.1 °C, Max=44.3 °C); Recipients (n=1350), (median temperature: 35.6 °C, quartiles: Q1=34.5 °C, Q3=36.6 °C, Min=30.0 °C, Max=42.9 °C); U=2249949, Z=-24.3545, p<0.001).

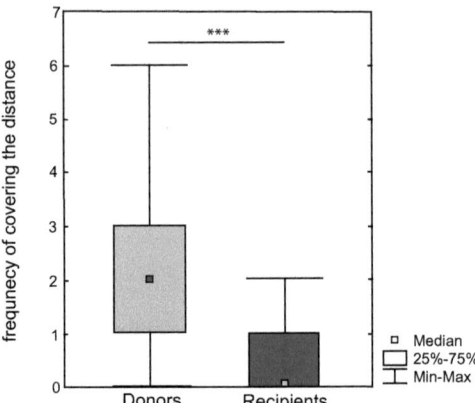

Fig. 4.3 Number of times (frequency) donors and recipients bridged the distance between brood comb and honeycomb

Donors covered the distance on average two times in one observation (Median=2, Q1=1, Q3=3, Min=0, Max=6), while recipients stayed on the brood comb (Median=0, Q1=0, Q3=1, Min=0, Max=2).
Mann-Whitney-U-Test:
n=46 (Donors) n=41 (Recipients) U=395.5, Z=4.65, p<0.000003

Trophallactic activities in the brood nest

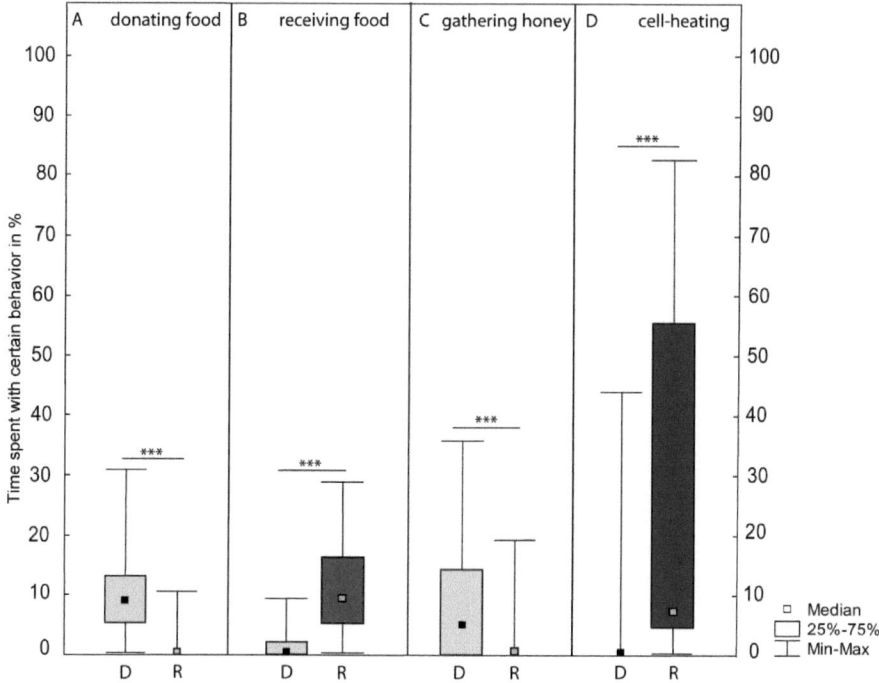

Fig 4.4 Behavioral patterns of donors and recipients
Time spent by donors and recipients in % of total observation time (light-grey = (D) Donors , dark-grey = (R) Recipients)

The donors D spent significantly more time with
(A) repeated donation of food (median: 8.9 %, quartiles: Q1=5.4 %, Q3=13.6 %, Min = 0 % , Max = 30.9 %)
(C) gathering of honey (median: 5.3 %, quartiles: Q=0 %, Q3=13.8 %, Min=0 %, Max=35.1%)
as the recipients R
(A) donation of food (median: 0 %, quartiles: Q1=0, Q3=2.5 %, Min=0 % , Max=10.7 %)
(C) gathering honey (median: 0 %, quartiles: Q1 and Q3=0 %, Min=0%, Max=18.2 %) did.

The recipients R spent significantly more time
(B) receiving food (median: 9.5 %, quartiles: Q1=5.4%, Q3=16.2%, Min=0% , Max=29.4%)
(D) cell-heating (median: 7.2%, quartiles: Q1=0 %, Q3=55.4 %, Min=0 % , Max=83.7 %).
The donors showed this behavior 0 % of the time
(B) receiving food (median: 0 %, quartiles: Q1=0 %, Q3=2.5 %, Min=0 %, Max=9.8 %)
(D) cell-heating (median: 0 %, quartiles: Q1 and Q3=0 %, Min=0 %, Max=44.0 %).

Mann-Whitney-U-Test: Donors (n=75), Recipients (n=75)
(A) donating food: U=185.0, Z=9.87, p<0.000001
(B) receiving food: U=311.5, Z=-9.40, p<0.000001
(C) gathering honey: U=1242.0, Z=5.90, p<0.000001
(D) cell-heating: U=1701.5, Z=-4.17, p<0.0003

Trophallactic activities in the brood nest

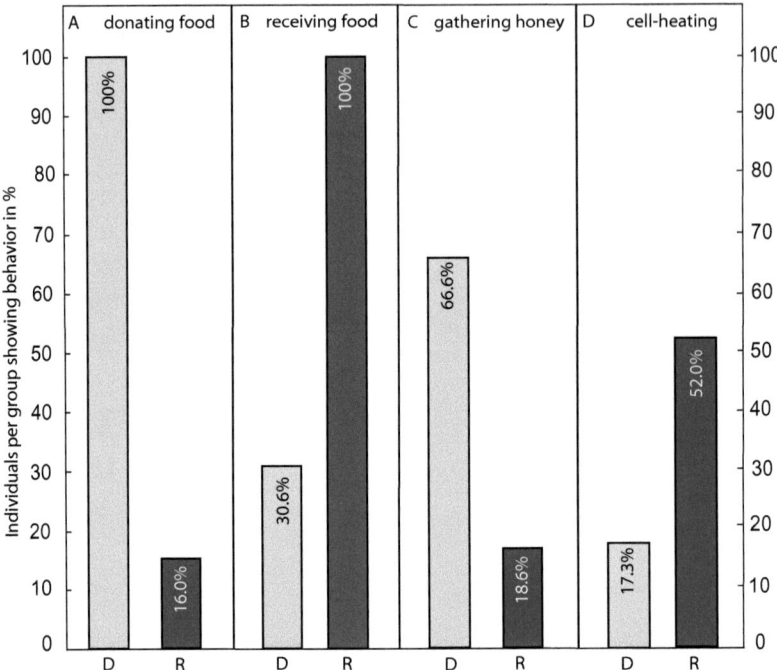

Fig. 4.5 Percentages of donors and recipients ever showing feeding or heating behavior

Absolute percentage of honeybees showing the behaviors A) - D) (light-grey = Donors D, dark-grey = Recipients R)
(A) Food donation was performed by 100 % of the observed donors and by 16.0 % of the recipients
(B) Food reception was performed by 100 % of the recipients and by 30.6 % of the donors
(C) Gathering honey was performed by 66.6 % of the observed donors and by 18.6 % of the recipients
(D) Cell-heating was performed by 52 % of the recipients and 17.3 % of the donors

Trophallactic activities in the brood nest

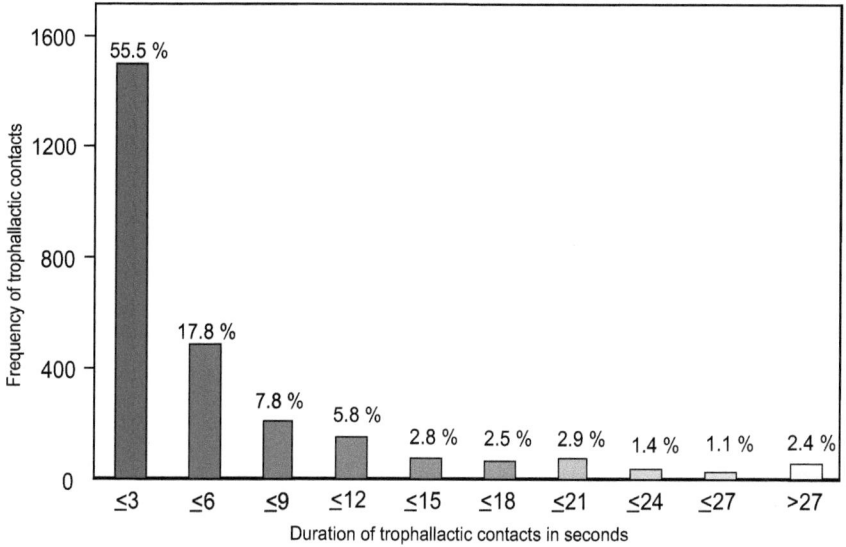

Fig 4.6 Feeding contact frequency and duration

Observed feeding contacts sorted by their duration. Most of the 2700 observed contacts lasted up to three seconds (55.5 %).
(Median: 3 seconds (quartiles: Q1=3 seconds, Q3=9 seconds, Min=1 second, Max=174 seconds).

5. Heat seeker: The honeybee feeding activity has a thermal trigger

5.1 Abstract

Honeybees use stored honey as fuel to heat the brood area of the hive and to keep the temperature in the winter cluster at an adequate level. In order to achieve this, bees metabolise sugar in their flight muscles to raise their body temperature to over 43 °C. The stored honey is predefined fuel for flying and heating, but is stored at a specific distance from the brood comb causing a potential logistic problem of efficient energy supply in the brood area.

Heating bees are supplied with honey by other worker bees which transport honey from the honeycomb to the brood area, ensuring that heater bees can then continue their work uninterrupted.

Here we demonstrate that trophallactic food exchange activity is closely related to the different parts of the hive entailing individual requirements for energy spent by heating bees.

Furthermore, we elucidate a heat-triggered mechanism that enables donor and recipient to accomplish food transfers without delay in spite of the total darkness of the hive in the brood area as the most energy consuming part of the hive.

Our results suggest a resource management strategy that has evolved from submissive appeasing behavior as it is seen in honeybees, bumblebees and other hymenopterans. This feedback mechanism reveals a new aspect of the division of labor and communication, and sheds new light on sociality in honeybees.

5.2 Introduction

Worker bees of *Apis mellifera* are capable of keeping the nest temperature at a certain range. During brood rearing, they maintain the temperature of their pupae between 33 °C and 36 °C by heating or cooling (HIMMER, 1927; SEELEY & HEINRICH, 1981; ESCH & GOLLER, 1991; HEINRICH, 1993) and in winter, they keep the core of the winter cluster between 20 °C and 35 °C (STABENTHEINER ET AL. 2003) since the nest periphery must not fall below the chill-coma temperature of 10 °C (FREE & SPENCER-BOOTH, 1960).

Another aspect of heat production in honeybees is aggression, for example during balling behavior (ESCH, 1960; ONO ET AL., 1987, 1995; STABENTHEINER, 1996; KASTBERGER & STACHL, 2003; KEN ET AL., 2005) or while attacking (ESCH, 1960; HEINRICH, 1971; ONO ET AL., 1987, 1995; STABENTHEINER, 1996; KASTBERGER & STACHL, 2003; KEN ET AL., 2005). Elevated body temperature is a signal for aggressive behavior prior to fighting or flight in many animal species. Especially in insects, where thoracic muscles need a certain "operating temperature", body heat is necessary to react to any kinds of stress threatening the individual or the colony. The interrelationship of aggressive behavior and thermoregulation was described precisely by STABENTHEINER ET AL. (2007). They found that guard bees, foragers, drones and queens were always endothermic, i.e. had their flight muscles activated, when involved in aggressive interactions. Especially guards make use of their endothermic capacity. Their mean thoracic temperature is 34.2 °C to 35.1 °C during examination of bees but higher during fights with wasps (37 °C) or attacks on humans (38.6 °C).

Heating and flying are highly energetic activities and consume the energy reserves of a worker bee rapidly (HEINRICH, 1993). Since honeybees use almost exclusively sugar as their energy substrate for muscular activity (JONGBLOED & WIERSMA, 1934; LOH & HERAN, 1970; SACKTOR, 1970; ROTHE & NACHTIGALL, 1989), the level of glycogen in the hemolymph must be kept high to provide an adequate fuel supply for the heat-generating flight muscles (CRAILSHEIM, 1988) which are among the most metabolically active tissues known (SOUTHWICK & HELDMAIER, 1987).

In honeybees, food can be stored in the crop or "honey stomach". Any liquid from the crop (nectar, water or honey) can be regurgitated and deposited in cells or transferred to other bees; therefore it is often referred to as "social stomach".

A crop-load of sugar solution can provide a bee with food for several hours. But even inactive, caged honeybees with a full crop held at room temperature die within 7 hours after being separated from their food source (HEINRICH, 1993). A physiologically challenging activity as flying or heating will consume their sugar fuel even faster, so the crop content and its sugar concentration consequently reflects the demand of the forthcoming task (NIXON & RIBBANDS, 1952; CRAILSHEIM, 1988).

During the brood rearing season, a typical comb is divided into four areas. The brood area where the heating activity takes place is usually situated in the center of the comb and surrounded by cells containing pollen. A circle of empty cells is enclosing the pollen cells.

The brood comb and the honeycomb are separated from each other by empty cells and the pollen circle. The empty cells situated between brood and food are kept empty as long as there is enough space for additional nectar deposits on the honeycomb and additional brood cells on the brood comb (Fig. 4.1, Fig. 5.1). The stored honey, which is the best source of carbohydrates in the hive, is kept in the upper corners of the comb (SEELEY & MORSE, 1976).

There are two ways how a trophallactic contact in honeybees can start: Firstly, a bee can beg for food by extending its proboscis and thrusting its tip towards the mouthparts of another bee (termed the donor if the contact leads to a transfer). If the donor bee responds by regurgitating food and thereby initiates a trophallactic contact, the begging bee was successful, and is termed the recipient. Secondly, a bee can offer food without being asked by opening its mandibles and moving its still-folded proboscis slightly downwards and forwards from its position of rest. A drop of regurgitated liquid food can be seen between the mandibles and on the proximal part of the proboscis (FREE, 1956). If a recipient bee touches that droplet with its antennae and then thrusts its proboscis between the mouthparts of the donor, this also results in a trophallactic contact (MONTAGNER & PAIN, 1971).

Offering food in a trophallactic contact is usually regarded as an appeasing or submissive gesture in hymenopterans (BUTLER & FREE, 1952; SAKAGAMI, 1954; MEYERHOFF, 1955; BREED ET AL., 1985; VAN DER BLOOM, 1991; WCISLO & GONZALES, 2006), because there is a correlation between individual worker dominance and trophallactic behavior (HILLESHEIM, 1989; KORST & VELTHUIS, 1982; LIN ET AL., 1999; HOOVER, 2006). Even though trophallactic contacts in honeybees usually appear to be conflict-free, aggressive behavior followed by trophallaxis can occur between honeybee workers as well. If, for example, foreign bees enter a colony or a bee is attacked in a cage experiment, the defeated or subdominant individual often regurgitates food (MONTAGNER & PAIN, 1971). Food offers as appeasing gestures is known in social and non-social insects as well as in mammals (Carpenter bees - VELTHUIS & GERLING, 1983; Hallictine bees – KUKUK & CROZIER, 1990; Social Vespidae - HUNT, 1991; Porine ants - LIEBIG ET AL., 1997; Bonobos -BLOUNT, 1990).

Trophallactic contacts occur all over the hive, but are carried out more frequently on the brood comb (SEELEY, 1982). As the food intake in honeybees is related to the task the worker is about to fulfil (NIXON & RIBBANDS, 1952; CRAILSHEIM, 1988) and the energetic requirements for the different activities in honeybees are unequal, the task partitioning system requires a sort of resource management that assures an ideal distribution of the available stocks to the worker bees that are performing the more strenuous activities.

Such a resource management system was described by BRANDSTETTER ET AL. (1988); they found that foragers get refuelled via trophallaxis between foraging flights by worker bees in the hive.

Recently, a similar mechanism to replenish the energy resources of the heating bees was described. BASILE ET AL. (2008) showed that heating bees stay on the brood comb and receive honey from cooler donor bees, which shuttle back and forth between honeycomb, where they collect honey from the cells, and brood comb, where they feed small portions to a number of heating bees.

However, the underlying motivation and the behavioral trigger for the donor's actions in both case remains unknown. If the recipient bees in the hive emit a certain signal and therefore initiate an underlying mechanism responsible for the behavior of the donors, the question arises, how this signal is produced, and how the donors decide to react properly.

5.3 Materials and Methods

All observations were made at the Beestation of Würzburg University from May 2005 to February 2007 using bees of the European subspecies *Apis mellifera carnica*.

We operated under red light conditions with four different two-frame observation hives, small arenas or with individual bees which were affixed in small plastic tubes. The ambient temperature of the room was controlled and kept at 20 °C ± 2. The worker bees were taken from 6 different colonies. These colonies were headed by unrelated artificially inseminated (10 to 12 drones) queens in order to provide genetic variance.

5.3.1 Behavioral observations in the hive

The behavioral observations were done at two-frame observation hives. Frames (internal diameter 5x5cm) were placed on the pane either over parts of the hive with defined cell content (summer) or over the center, the periphery of the winter cluster (winter) respectively. The observed area was small enough to be monitored by a single person without missing a feeding contact, but still contained enough bees to observe several feeding contacts in every run (number of total runs: n=226).

We measured the trophallactic activity in parts of the hive depending of the cell contents in the area beneath the bees. We defined the trophallactic index as the number of trophallactic interactions per 15min, divided by the average number of bees present (the number of bees was recorded every five minutes i.e. three times for every run). After 15min, the frame was moved to another part of the hive.

In winter, the comb is no longer divided into the four areas as under summer conditions, because there usually is no brood, and therefore no brood-related heating either.

We defined the areas according to the heating behavior in the winter cluster. One frame was placed over the hottest part of the winter cluster (usually close to the center), and the other frame was placed in the periphery of the cluster where less heat is produced but bees linger anyway. The temperature difference between the two areas had a mean difference of Δ 3.4 °C (SD: Δ 1.8 °C) and was measured before and during observations with an automated measurement instrument with temperature sensors (ALMEMO 2290-8 V5).

We replaced the panes of the observation hives with heat permeable foil and took close-up shots of the brood comb using a thermal imaging camera (S40 from FLIR Systems $^{TM)}$. We used the plastic wrap "Cling Wrap"© made of 100% polyethylene manufactured by the U.S. American company GLAD™. Since the foil still alters the radiation emanating from the object, we needed to define the error produced at different temperatures. The error produced by the foil is non-linear, therefore every measured thoracic temperature had to be corrected specifically for every picture taken (see chapter 3, Suppl. Fig. 2. and Suppl. Tab. 2).

The film footage consists of 22h (88 sequences à 15min) taken on 11 different days (May and June 2005) between 11:00a.m. and 14:00p.m. After each sequence, the camera was moved to another part of the brood comb or to another observation hive in order to avoid pseudoreplicates. The recorded images were analysed with the software ThermaCam ^{TM} Researcher Pro2.7 (FLIR Systems $^{TM)}$.

The use of a thermal camera was necessary because not all trophallactic contacts lead to food transmission. With a thermal imaging camera it is possible to record the behavior as well as the transmission of food, because the liquid food either cools down or warms up the proboscis of the participants and therefore changes the contrast in the thermal image (BASILE ET AL., 2008).

In addition, we measured the angle at which the bees are positioned to each other before and during food transfer. The position of the body axes of donors and recipients were noted one second before the contact (=body axis before trophallaxis) by drawing them as straight lines over the bees and during trophallaxis (=body axis during trophallaxis) by drawing a second set of straight lines. An angle results between the two straight lines and illustrates the body axes before and during trophallaxis of each participant, and displays the shift of body axes of donor and recipient respectively. This shift originates in the movement each bee makes in order to reach the mouthparts of the other participant (Fig. 5.2).

5.3.2 Warm-up experiments in the hive

We chose the same set up as for the observation of the trophallactic index in winter. One observation frame was placed onto the hottest spot of the winter cluster (24 °C to 30 °C) and the second frame in a cool (18 °C to 22 °C) part of the cluster. The hottest spot of the winter cluster was detected with an automated measurement instrument with temperature sensors (ALMEMO 2290-8 V5) (the temperature was recorded every five minutes i.e. three times for every run). Observations were performed as previously.

In the next step the whole hive was isolated apart from the area of the cool frame. This particular part was then heated up to approximately 35 °C by an infrared lamp. The temperature of the frame was kept under control by moving a lamp closer to the pane or further away. The temperature was monitored with an automated measurement instrument with temperature sensors (ALMEMO 2290-8 V5) and ensured that it was stable at 35 °C. This made it possible to warm up the cool part of the winter cluster and measure trophallactic activities in that artificially warmed up environment and compare it to the naturally heated and the formerly cool part of the cluster.

We recorded the experiment and analysed the video footage afterwards, since the offering behavior we measured in the warm-up experiments is carried out quickly and therefore is hard to count without video footage (digital video camera, DCR-SR 190 E Sony). The footage was taken on 5 consecutive days (February 2007) and consists of 7.5h (30 sequences à 15min) and the trophallactic index was calculated as well as the offering index (total number of offers divided by the average number of bees present).

5.3.3 Warm-up experiments in the arena

To observe the impact of heat on trophallactic behavior in an even more temperature controlled environment, we conducted experiments with individual bees in an arena in May and June 2005. The bees were randomly taken from one hive at a time. Since the behavior we were observing is related to brood, only workers from the brood area were used for this experiment.

Four bees were killed by deep-freezing and were each fitted with a carbon film resistor in the thorax. Then the bees were arranged in a circle. The experiments were conducted in a 20x20cm duroplastic arena covered with a glass pane. The resistors´ wire ends were connected one at a time to a transformer (Amrel Linear Power Supply LPS-301) with an output of a constant voltage of 4V at 0.04A warming up the resistor to 40 °C.

Each was connected to the transistor once for five minutes. Thus we could create a situation where we were able to control the heat emission i.e. thoracic temperature of the dead bees, and observe the behavior of living bees towards the warm or cold nestmates. We measured the frequency of occurrence of offering and begging behavior towards dead worker bees. After heating every bee once, the arena was cleaned and all bees were replaced by a new group of one living and four dead workers.

5.3.4 Warm-up experiments with restrained bees

The worker bees were randomly taken from one hive at a time. Since the behavior we were observing is related to brood care only, workers from the brood area were used for this experiment. The experiments were conducted in May and June 2006.

Bees were restrained in small plastic tubes so they could only move their heads, mouthparts and antennae freely a method similar to that used in proboscis extension response measurements (BITTERMAN ET AL., 1983). The transformer and the output we used were used in the same way as in the arena experiment. We utilised two types of resistors: carbon film resistors (for heat emission) and high power wired wound resistors (for emission of an electromagnetic field).

Each restrained bee was tested ten times for one minute on its behavior towards the cold and one minute towards the warm resistor.

The observed behaviors were: offering food, i.e. regurgitating a droplet of liquid food between mandibles and proboscis, begging for food, i.e. a proboscis extension, "snatching", i.e. spreading and closing the mandibles, and showing none of the mentioned behaviors at all.

The "snatching" behavior we observed is described as an aggressive behavior in honeybees (SAKAGAMI, 1954), ants (CROSLAND, 1990), eusocial wasps (O´DONNELL, 2001) and other insects (ALEXANDER, 1961).

Since honeybees are known to respond to electromagnetic fields, (GOULD, ET AL., 1978, 1980; GOULD, 1980) which is produced by the resistors as well. In order to confirm that the emitted heat and not the electromagnetic field is the trigger for the observed behavior we conducted additional runs with a high power wired wound resistors These resistors produce an electromagnetic field, but do not heat up due to their higher resistance. Additional worker bees (n=51) were tested five times for reactions toward each of the two resistor types.

All statistical analyses were performed using the statistical package Statistica 8©.

5.4 Results

5.4.1 Trophallactic behavior in the observation hives

During summer, trophallactic contacts were observed in every area of the hive but with different frequencies depending on the cell contents. The trophallactic index was highest near the center of the comb on brood and pollen cells and declined on empty cells and on the honeycomb. The differences between the trophallactic index of the capped brood area and the empty cell area, and between the capped brood area and honey cell area were highly significant. Differences between the trophallactic indices of the pollen cell area and the empty cell area, and the pollen cell area and honey cell area were highly significant as well.

Even though the trophallactic index was relatively high on capped brood cells, there was no significant difference to the index on pollen cells. Likewise, the trophallactic index of empty cells and on honey cells was not significantly different (Fig. 5.1).

Similar results were obtained in winter. The trophallactic index in the warm area of the cluster significantly exceeded the index of the cool area (Fig. 5.3).

The temperature within the frame and the trophallactic index correlated significantly and positively (Spearman coefficient: n=112, R=3.7, p<0.05).

5.4.2 Feeding contacts

A vast majority of feeding contacts on the capped brood (93 % of n=759) were induced by an offering bee. Offering worker bees visibly regurgitated food and receiving bees touched the food droplet with one antenna, thereby releasing a proboscis extension response (Fig. 4.3A). The recipient's proboscis was then thrusted between the mouthparts of the offering bee and the food was transferred. Significantly less (only 7 %) of the trophallactic contacts were induced by a recipient bee, showing begging behavior by trying to thrust the proboscis between the mouthparts of a potential donor bee which regurgitated food as consequence. (X^2-Test: n=759, X^2=1324.5, p<0.001).

Not only were most trophallactic contacts initiated by the donors, they also shifted their bodies more vigorously than the recipients in order to access the contact (Fig. 5.2A, B). The angle we measured between the body axes a second before and the body axes while food was being transferred was significantly greater in donors as in recipients (Fig. 5.2C).

5.4.3 Winter with artificial heating

The trophallactic indices of areas heated by the workers or by the artificial source were not significantly different (Fig. 5.3). The trophallactic index between the naturally warm area in the center of the cluster and the cooler area in the periphery of the cluster showed a significant difference, which confirms the observations we made without video support. The offering index showed no significant difference between the warm and the cool area.

The trophallactic index in the artificially warmed area was significantly higher than in the same area before warm up. There was no significant difference between the trophallactic index in the naturally warm and the artificially warmed up areas (Fig. 5.4A).

The offering index in the artificially warmed area was significantly higher than before the warm up. It was also significantly higher than the offering index in the naturally warm area (Fig. 5.4B).

5.4.4 Area experiments

Some honeybees in the arena reacted to their dead hive mates with food offering or begging. Offering behavior was shown significantly more often towards a hot dead bee. (X^2-Test: n=39, X^2=27.1, p<0.001)

Begging behavior occurred rarely in general and occurred significantly less frequent towards the cold bee (Fig. 5.5).

5.4.5 Reactions to heat and electromagnetic fields

Honeybees reacted more frequently towards the warm carbon film resistor. They offered food more often towards the warm resistor and if they begged for food, they begged mostly from the warm resistor. Bees reacted with snatching significantly more often towards the warm resistor. Those bees that were showing a particular behavior did so significantly more in presence of the cold resistor (Fig. 5.6, Tab. 5.4).

In an additional experiment, we tested the high power wired wound resistors against carbon film resistors. The bees reacted less frequently towards the high power wired wound resistors compared to the carbon film resistors in general. Offering food and snatching was shown more often towards the carbon film resistor. Begging was shown infrequently and with no significant difference between both resistors. Furthermore, bees that showed no reaction at all did that more frequently towards the high power wired wound resistor (Fig. 5.7, Tab. 5.5).

5.5 Discussion

The feeding activity (the trophallactic index) in the hive proved to be highest on the brood comb and the surrounding pollen circle in the brood-rearing season and the core of the winter cluster in the brood-free cold season. Similar results during brood-rearing have been described by SEELEY (1982), who observed trophallactic activity all over the hive but particularly often on the brood comb.

The measured increase could be caused by various factors. Firstly, it cannot be an artifact of the overall activity since our index controls for that.

Secondly, the increase in trophallactic activity could be a side effect of the nurse bees´ feeding activity on the brood comb. Nurse bees mainly provide the larvae with proteinaceous jelly, but also supply drones and adult bees with it.

However, BASILE ET AL. (2008) showed that bees which acted as donors in trophallactic contacts on the brood comb constantly shuttle between honeycomb, where they gather honey, and brood comb, where they feed heating worker bees. The idea that donor bees are in fact nurse bees is not supported by this data, because neither donors nor recipients engaged in larval feeding activity.

Thirdly, the most noticeable similarity between the brood comb and in the core of the winter cluster is the heating activity, which might be the major cause for the increased trophallactic index. A close connection between the frequency of feeding contacts and brood was described by ISTOMINA-TSVETKOVA (1958). She found that when brood rearing in autumn was reduced, there were less feeding contacts in the hive. Nevertheless, the presence of brood alone is most unlikely to be the trigger for an increase or decrease for the trophallactic index, since we measured a similar ratio of feeding activity between the warm core of the winter cluster and the cooler periphery, where there was no brood but heating activity. The fact that the temperature and the trophallactic index within the frame correlated significantly and positively indicates that the more the bees heated, the higher was the feeding activity in that particular area. In addition, the observation of BASILE ET AL. (2008) that donors feed recipients with an elevated thoracic temperature on the brood comb supports the conclusion that heat is the main trigger for the increased trophallactic index on brood.

Indeed, that conclusion is supported by the results of the artificial heating experiments at the periphery of the winter cluster. By artificially heating up the area to the normal levels of the core of the winter cluster, we increase the feeding activity to the same high level which are found in the warmer core area (Fig. 5.4A).

Apart from an increase in the trophallactic index by artificial heating, we found a significant increase in food offering behavior in the formerly cold area after warming it up to 35 °C. Since there was no difference between the offering behavior between the cold periphery and the naturally warm part, it raises the question how offering, trophallactic behavior and heat are connected?

Offering as a first step in honeybees' trophallactic contacts can easily develop into feeding, if the recipient is willing to accept the offer. The artificially warmed part has a high offering index as well; therefore, if enough bees are willing to take the offer, a high trophallactic index is consequential.

But why has the naturally warm area a lower offering index than the artificially warmed area?

Offering almost instantly leads to a trophallactic contact if there is need for food, so the naturally warmed area has a high level of feeding activity due to energy expenditure. Within these areas there is a high demand and high supply. Within the artificially heated area heating is done not by the bees which results in a lower demand for energy and therefore results in a reduced PER responses of workers to food offerings. They have no real need for sugar and therefore are less receptive than high energy spending heating bees.

Another aspect of the high offering index is that in this experiment the food is not offered to relatively small area (the sum of all single heating bees) as in the naturally warm area in the center of the winter cluster, but to a proportionally much larger warm surface (heated pane), which might be another cause for the increased offering.

Artificial warm-up experiments with honeybees were also conducted by STARKS and GILLEY (1999). They heated a larger area of the hive in brood-rearing season in order to observe heat-shielding behavior. This heat-shielding is performed mostly by younger bees and is a mechanism to protect the brood from temperatures above 36 °C. Heat-shielding bees turn their ventral side towards the heat source and sometimes even spread fluid in order to lower temperature by evaporation. The bees in our experiment are unlikely to having shown heat-shielding behavior. Firstly, the temperatures we exposed a small part of the hive to are similar to the upper temperatures in the core of the winter cluster. Even if some bees regurgitated fluid, other bees react with a PER and absorb the fluid indicating that the fluid is from the crop and that it is food and not liquid to cool off. Furthermore, the liquid was never spread over the head or the thorax of the bee nor was it spread over any area of the hive.

Secondly, since there was no brood to be protected in the hive while conducting our experiments, it is unlikely to observe heat-shielding behaviour since it is a task which is directly connected to the protection of the brood.

In addition, we showed how closely the trophallactic index and the offering behavior are related to one another by analysing the initiation of the trophallactic contacts on the brood comb. Most feeding contacts were initiated by the donor, which was offering food to the recipient, and not initiated by the recipient begging for food. This behavior carries an important economical advantage for the honeybees: If a trophallactic contact is initiated by a recipient, it has to beg for food until it finds a donor willing to regurgitate food. This action takes at least three steps: (1) the recipient begs for food by extending its proboscis and trying to reach the mouthparts of a potential donor. (2) The potential donor either decides to regurgitate food or not. (3) If the donor regurgitates food, trophallactic contact can be established. If the potential donor refuses to regurgitate food, the recipient has to re-orientate itself and steps 1 to 3 must be repeated until a suitable donor is found. This means that the success of the trophallactic contact depends solely on the action of the potential donor.

If, by contrast, the trophallactic contact is initiated by the donor, the initiation of the trophallactic contact takes only two steps: (1) the donor offers food by regurgitation. (2) If a bee touches the droplet of food between the spread mandibles of the donor with its antennae and has a low threshold for sugar (KUWABARA, 1957), a PER is triggered causing its proboscis to extend directly towards the food and the contact can be established (Fig 5.8).

The threshold for a PER is closely related to the energy status of a bee. Therefore, a heating bee is more likely to have a low blood sugar status and react to the sugar stimulus than an inactive bee. More precisely, the offered food is more likely to be accepted from bees in need of food. The antennae play a major role in releasing both offering and begging behavior. The importance of antennal contact in releasing the food transfer and in helping bees to orientate the mouthparts to one another was recognized by FREE (1956). He found that worker bees with no antennae were less often offered food than those with both antennae intact. In addition, he found that workers from which both antennae had been completely or partially removed, did not beg for food to any extent. The loss of the antennae might have the consequence of not being able to perform a PER and consequently not being able to accomplish a trophallactic contact.

Even though the PER in honeybees is an established tool for various learning experiments, its primary function remains unknown. Our findings show a possible purpose of the reflex, in supporting the orientation towards a food source in the darkness of the hive.

The offering of food to the recipients has another economical effect: As BASILE ET AL. (2008) described, the donors shuttle between honeycomb and brood comb, while the recipients stay on the brood comb and are able to continue the heating task virtually uninterrupted. If the food offered by those donors is gathered honey of high sugar content, it is most likely to trigger a PER. The donors´ activity not only consists of the transportation of honey to the area where food is needed, but of offering it to the individual bee in need.

Such a system would not only be time-saving in avoiding the begging and potential refusing of regurgitation, but energy-saving as well. The abundance of recipients on the brood comb increases brood rearing efficiency (SCHOLZE, 1964; SOUTHWICK & HELDMAIER, 1987; CRAILSHEIM, 1988) and improves the efficiency of insulation against heat loss (SACKTOR, 1970, KRONENBERG & HELLER, 1982; SOUTHWICK & HELDMAIER, 1987). A similar activity of donors refueling heat-emitting bees has been described by BRANDSTETTER ET AL. (1988). The feeding of foragers in the hive might be work on the same principle, since foragers attain elevated thoracic temperatures in the hive before they leave for their flight.

Furthermore, the results of the axis shifts supports that donors are the more active part in that interaction by showing a stronger shift in the body axis towards the recipients before the contact.

Honeybee communication in the hive excludes visual cues to a large extent, but includes cues like vibration as in dance communication and smell as in kin recognition.

FREE (1956) showed that feeding behavior is closely related to smell. He tested honeybees for offering and begging behavior towards dead hive mates and dummies, with and without the hive mates' odor. The bees reacted by performing offering and begging behavior most strongly towards dead bees and dummies with their hive mates´ odor.

We used a similar set up for our experiments, but added heat as a potential eliciting factor for feeding behavior.

Begging was shown towards warm or cold bees with the same infrequence, but the warm bees were offered food significantly more often than the cold bees. This is not surprising, as hot honeybees emit increased odor, due to the fugacity of volatile hydrocarbon compounds in their cuticle. This connection between heat and odor is known to play a major role in hive mate recognition by the guard bees. STABENTHEINER ET AL. (2002) found that when guards examine returning foragers, the examinee often increases its thoracic temperature to enhance chemical signalling during examinations. Therefore, body temperature is a parameter that has to be considered in research on nestmate recognition in research on trophallactic behavior in the honeybee. Accordingly, our results could be a reflection of the stronger odor or a reaction to heat emission.

To exclude odor as cue, we tested individual bees on their feeding reaction towards a warm small wire resistor. In this setup the bees were no longer able to move freely and choose a certain distance to the heat emitting source. The reactions towards the warm resistor were similar concerning the feeding reaction. The bees offered the warm resistor significantly more often food than the cold resistor. Begging behavior was shown with similar infrequence towards the warm and the cold resistor.

A repetition of the experiment with high power wire wound resistors showed that worker bees do react to the electromagnetic field with aggression and begging, but less frequently than to the small wire resistor and offering behavior was almost never shown at all. Therefore, we can conclude that the offering behavior of the bees was primarily influenced by the heat and not by the electromagnetic field emitted by the resistors.

However, the reactions of the honeybees were broadened by an aggressive behavior towards the heat source in this experiment. The spreading of the mandibles is regarded as an aggressive behavior in honeybees (SAKAGAMI, 1954) and since the bees could only move their heads and mouthparts it was the only way of showing aggression at all in this experimental set-up.

The remaining question is: why do donors tend to regurgitate food in the presence of heat? Bees are known to regurgitate liquid and to spread it over the surface of their heads and even on their thoraces to cool off if threatened by overheating.

In our experiments, the temperature affecting the bees emitted from the resistor (40 °C) and the pane heated by the infrared lamp (36 °C), never reached critical heat. The temperatures under which we conducted our experiments are the same as those honeybees emit before takeoff, while heating the winter cluster, or while preparing for an attack (HIMMER, 1927; ESCH, 1960; HEINRICH, 1971; SEELEY & HEINRICH, 1981; ESCH & GOLLER, 1991; HEINRICH, 1993; ONO ET AL., 1987, 1995; STABENTHEINER, 1996; STABENTHEINER ET AL. 2003; KASTBERGER & STACHL, 2003; KEN ET AL., 2005).

Honeybees are also known for regurgitating food as an appeasing gesture to avoid conflicts. The connection between heat emission, which is a sign for aggression (DALTON, 1940) and food offering, which is an appeasing gesture, might be a main step in the process of signal evolution. If a behavior becomes stereotyped and changes function, it is referred to as ritualized (TINBERGEN, 1952). It is generally recognized that ritualization has played a major role in the evolution of communication in social insects (HÖLLDOBLER, 1984; WILSON, 1985B) especially in the honeybees´ dance communication. Only recently, RITTSCHOF and SEELEY (2008) presented a case of ritualization in the honeybee´s buzz-run.

The process of signal evolution is described as a procedure in which the establishment of an association between the particular condition of the sender and the production of the cue by the sender is the first step (OTTE, 1974). In our case, the condition of heating and the cue "heat" are easily associated with one another, since the cue is an inevitable byproduct of heating.

The second step in the signal evolution is for potential receivers to detect the cue and to use its occurrence to improve their decision making. Honeybees are able to discriminate temperatures of at least Δ 0.25 °C (HERAN, 1952) and heat is a noticeable cue for aggression which is often answered with food offering. In order to establish a stable feedback in this ritual, the receivers´ decision-making must boost the fitness of the sender, which will improve the detectability of the cue. In our example, the decision to offer food does boost the fitness of the sender and enhances the intensity of the cue itself by rewarding the sender with a food offering as a submissive gesture and a positive one by fueling the heating process with the necessary sugar load.

The benefit of this ritualized appeasing gesture is evident, since the offering behavior is target-oriented towards energy-consuming hot bees. Every hot individual contributes to the welfare of the hive either by heating the brood, keeping the winter cluster warm, by heating up for a foraging flight or by raising the body temperature to attack an intruder. In addition, the offering is only accepted, if the threshold for sugar is low enough in the recipient.

Comparing this elaborate energy distributing system with related groups, like bumblebees, or social wasps, leads to the conclusion that there is no equivalent system. In other hymenopteran societies, brood heating is either not regulated strictly, or there is no trophallaxis between the adult members of a colony in non-apine hymenopterans. Apart from that, a winter cluster is a phenomenon solely formed in honeybees.

Only the honeybee has evolved to a point that enables the colony perennial survival and provides the opportunity to produce the next generation of workers as soon as spring provides enough pollen.

This system of providing the heat-emitting bees with honey by using their heat as a cue, and using a formerly appeasing gesture as ritualized response contributes to the highly economical resource management that is in line with the ecological requirements in the colony and at the same time with the physiological conditions of the individual.

5.6 Appendix – Figures and Tables

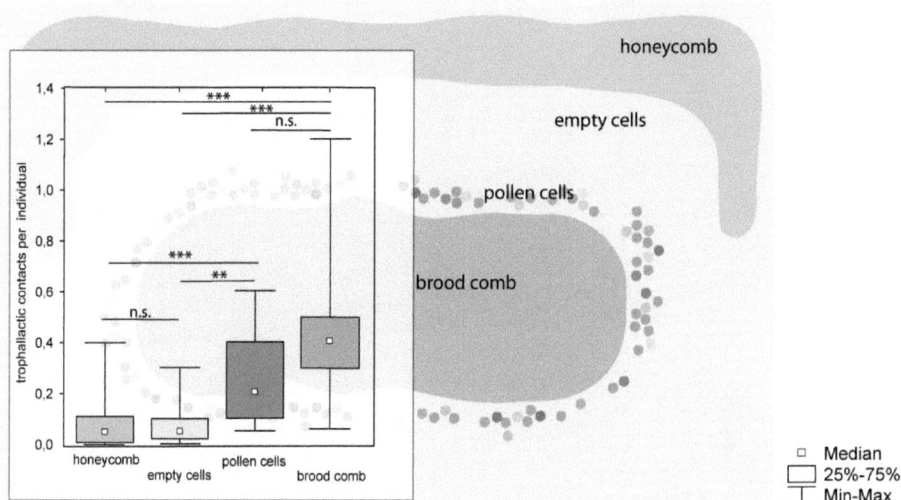

Fig. 5.1
Schematized comb with four different areas and boxplot with trophallactic indices

Trophallactic contacts per individual (Trophallactic contacts/mean number of bees) within 15 minutes and p-values. (Kruskal Wallis Anova : n_{total}=226) For details and statistics see Tab. 5.1.

Heat seeker

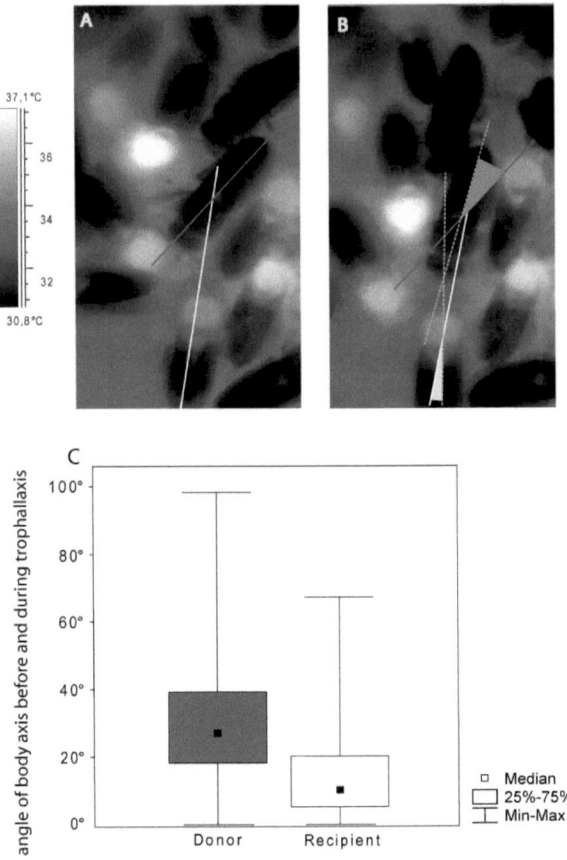

Fig. 5.2 Shift in body axis of donor and recipient before and during trophallaxis

Recipient (grey line) and donor (white line) move and therefore their body axis shift from A) before trophallaxis (solid line) and B) during trophallaxis (dotted line). The donor moves in a larger angle (grey triangle), the recipient moves less (white triangle).

A) one second before trophallactic contact
B) during trophallaxis
C) Angle of body axis before and during trophallaxis in donors and recipients

Mann-Whitney-U-test: n_{Donor}=121, median: 27, Q1=18, Q3=29, min=0, max=98; $n_{Recipient}$=121, median:10, Q1=5, Q3=20, min=0, max=67; U=2836.0, Z=8.24, p<0.001

Heat seeker

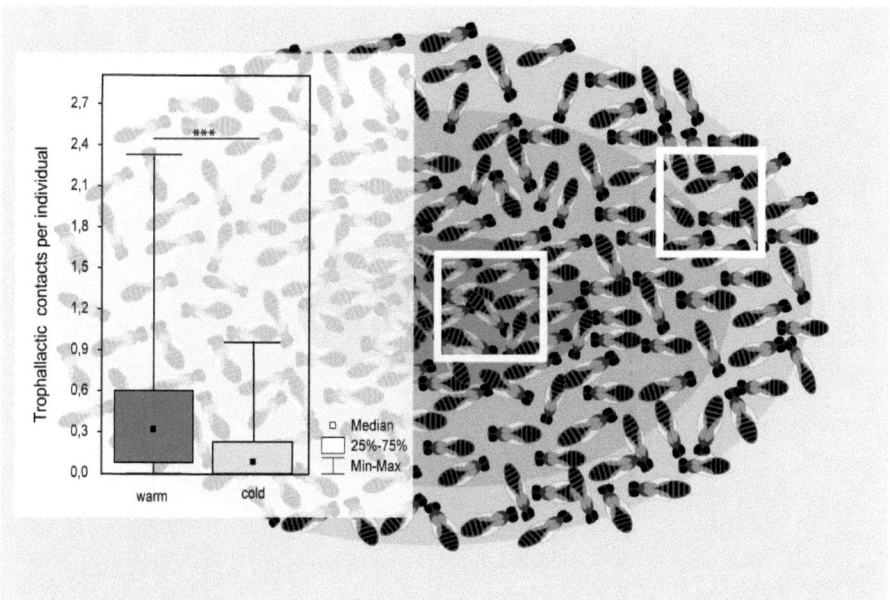

Fig. 5.3 Schematized comb with winter cluster and boxplot of warm and cold frame trophallactic indices

Mann-Whitney-U-test: nwarm=56, median: 0.2, Q1=01, Q3=0.45, min=0, max=1.7; ncold=56, median: 0, Q1=0, Q3=0.2, min=0, max=0.8; U=840.5, Z=4.23, p<0.001

Heat seeker

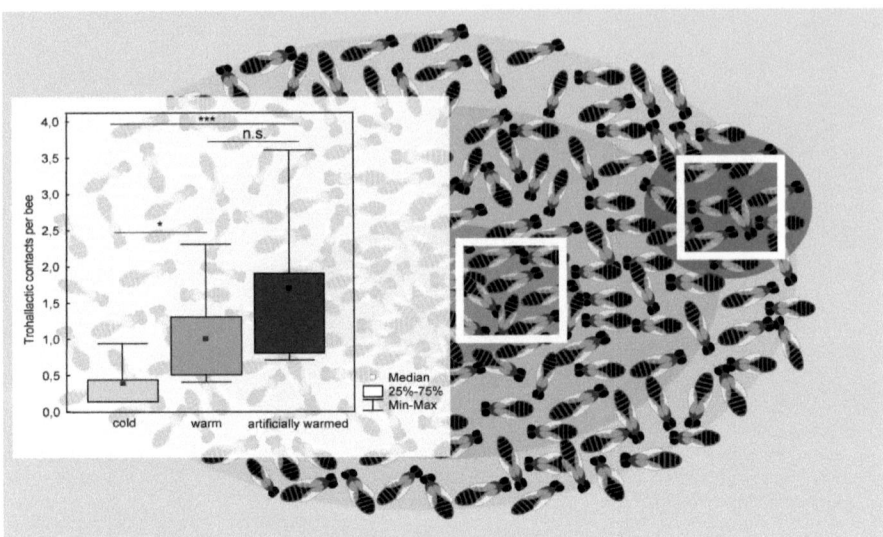

Fig 5.4 A Schematized comb with winter cluster and boxplot of naturally warm center, cold and warmed up frame trophallactic indices
Trophallactic contacts per individual (Trophallactic contacts/mean number of bees) within 15min and p-values. (Kruskal-Wallis Anova: n_{total}=90) For details and statistics see Tab 5.2

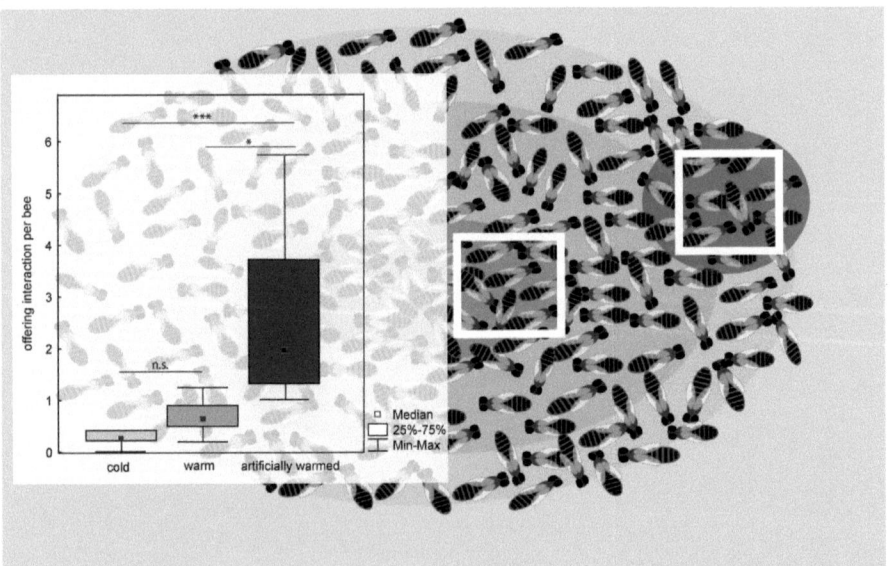

Fig 5.4 B Schematized comb with winter cluster and boxplot of naturally warm center, cold and warmed up frame offering indices
Offering behavior per individual (Offering behavior/mean number of bees) within 15min and p-values. (Kruskal-Wallis Anova: n_{total}=90) For details and statistics see Tab 5.3

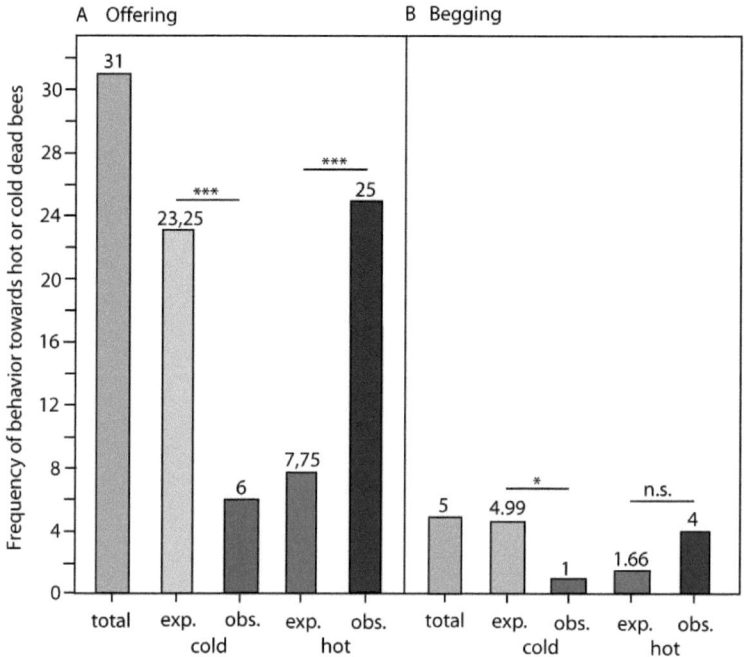

Fig. 5.5 Offering and begging behavior towards hot or cold dead bees

Observed vs. expected test with p-values: $n_{offering}=31$, cold: expected=13.25, observed=6, $p<0.001$; hot: expected=7.75, observed= 25, $p<0.001$; $n_{begging} = 5$, cold: expected=4.99, observed=1, $p<0.03$; hot: expected=1.66, observed= 4, n.s.

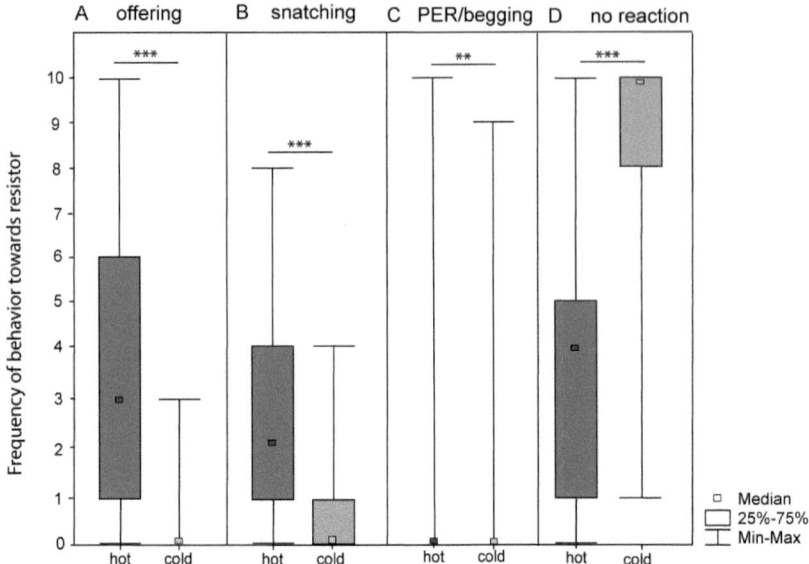

Fig. 5.6 Frequency of behavior of restrained bees towards a warm and cold carbon film resistor
(Wilcoxon-matched pair test: n=56) For details and statistics see Tab. 5.3.

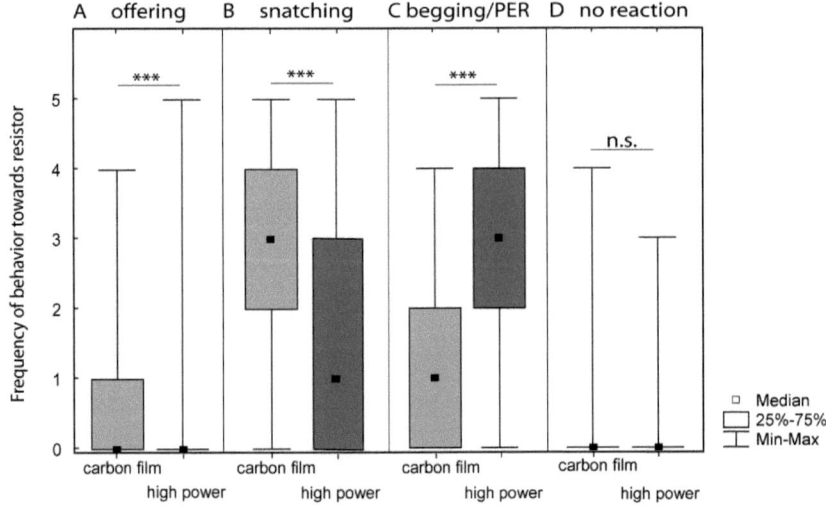

Fig. 5.7 Frequency of behavior towards carbon film and high power wire wound resistors
(Wilcoxon-matched pair test: n=51) For details and statistics see Tab. 5.4.

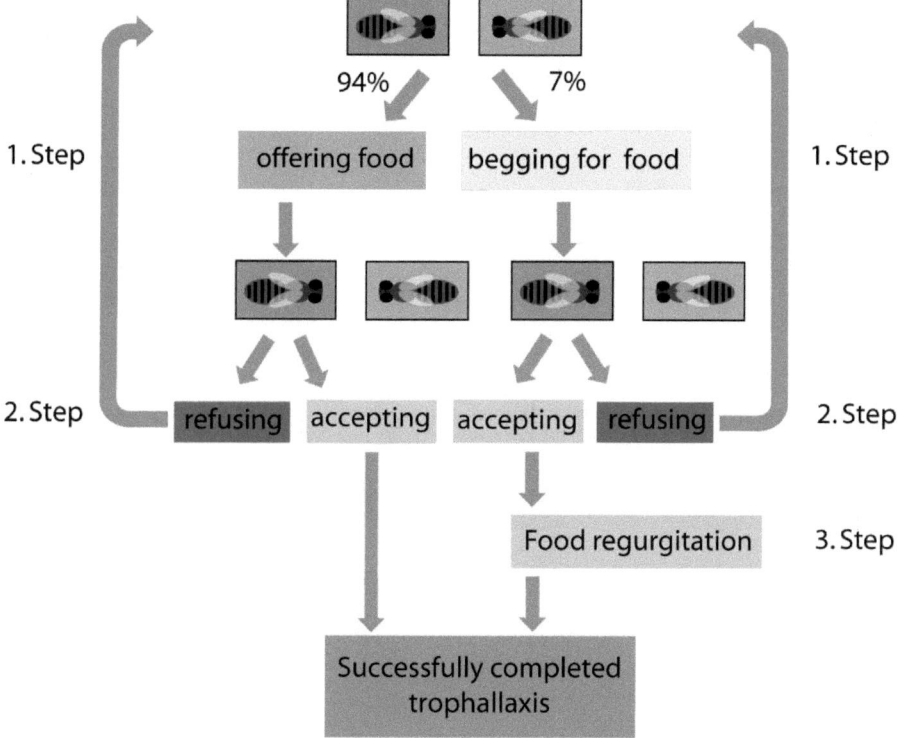

Fig. 5.8 Trophallactic interaction of donor and recipient on the brood comb

Depending on the initation of a trophallactic contact it takes 2 or more steps to accomplish a food transfer. Offering or begging is step 1. The decision whether to accept or refuse the offer or the begging is step 2. If a food offer is accepted, the food is transfered after two steps. If the begging is accepted, the bee first has to regurgitate food (step 3) before the food can be transered. If the bee decides to refuse the food offer, the donor has to reorientate and try to offer the food to another bee (Repetition of step 1 and 2). If a bee refuses to reguritate food for a begging bee, the potential recipient has to reorientate and try to begg from another bee (Repetition of step 1 and 2).

Since a food offer already includes reguritation, a trophallactic interaction that starts with a food offer depends mostly on the threshold for sugar in the recipient. The delay between accepting the offer and trophallaxis is very brief, since the offered food releases a PER.

If the begging bee is successful, the donor bee has to regurgitate the food in order to make trophallaxis possible. The decision mostly depends on the fill level of the crop and even if sucessfull, the operation takes more steps and therefore more time.

Heat seeker

	Honeycomb (n=49)	Empty cells (n=72)	Pollen cells (n=15)	Brood comb (n=90)
Min	0.0	0.01	0.07	0.08
Q1	0.02	0.03	0.12	0.3
Median	0.06	0.07	0.21	0.42
Q3	0.12	0.11	0.41	0.51
Max	0.4	0.3	0.6	1.2

	Honeycomb (n=49)	Empty cells (n=72)	Pollen cells (n=15)	Brood comb (n=90)
Honeycomb	-	n.s.	p<0.001	p<0.0000001
Empty cells		-	p<0.002	P<0.0000001
Pollen cells			-	n.s.
Brood comb				-

Tab. 5.1 Trophallactic indices on different areas of the comb (summer)

(Kruskal-Wallis-Anova and multiple comparisons of all groups for exact p-values)

	Heat seeker		
	Cold (n=30)	Naturally warm (n=30)	Artificially warmed-up (n=30)
Min	0.17	0.42	0.75
Q1	0.17	0.51	0.86
Median	0.44	1.1	1.76
Q3	0.44	1.34	1.96
Max	0.97	2.39	3.63

	Cold (n=30)	Naturally warm (n=30)	Artificially warmed-up (n=30)
Cold	-	n.s.	p<0.0001
Naturally warm		-	p<0.04
Artficially warmed-up			-

Tab. 5.2 Trophallactic indices on different areas of the winter cluster

(Kruskal-Wallis-Anova and multiple comparisons of all groups for exact p-values)

Heat seeker	Cold (n=30)	Naturally warm (n=30)	Artificially warmed-up (n=30)
Min	0.0	0.3	1.0
Q1	0.3	0.5	1.35
Median	0.3	0.6	1.95
Q3	0.45	0.85	3.7
Max	0.45	1.3	5.8

	Cold (n=30)	Naturally warm (n=30)	Artificially warmed-up (n=30)
Cold	-	n.s.	$p<0.0001$
Naturally warm		-	$p<0.04$
Artficially warmed-up			-

Tab. 5.3 Offering indices on different areas of the winter cluster

(Kruskal-Wallis-Anova and multiple comparisons of all groups for exact p-values)

	Offering T=0.0, Z=6.4		Snatching T=6.1, Z=5.2		Begging T=9.0, Z=2.4		No reaction T=0.0, Z=6.4	
Resistor	hot	cold	hot	cold	hot	cold	hot	cold
Min	0	0	0	0	0	0	0	1
Q1	1	0	0	0	1	0	1	8
Median	3	0	0	0	2	1	4	10
Q3	6	0	0	0	4	1	5	10
Max	10	3	10	9	8	4	10	10
p-value	$p<0.000001$		$p<0.0000001$		$p<0.01$		$p<0.0000001$	

Tab. 5.4 List of all observed behaviors towards hot and cold carbon film resistors and p-values (Wilcoxon-matched pair test: n=56).

	Offering T=6.0, Z=3.2		Snatching T=87.0, Z=4.7		Begging T=15.0, Z=1.3		No reaction T=40.5, Z=5.2	
Resistor	hot	cold	hot	cold	hot	cold	hot	cold
Min	0	0	0	0	0	0	0	0
Q1	0	0	2	0	0	2	0	0
Median	0	0	3	1	1	3	0	0
Q3	1	0	4	3	2	4	0	0
Max	4	5	5	5	4	5	4	3
p-value	$p<0.001$		$p<0.000003$		n.s.		$p<0.01$	

Tab. 5.5 List of all observed behaviors towards carbon film and high power resistors and p-values (Wilcoxon-matched pair test: n=51).

6. General discussion

The ecological success of social insects is largely based on the complex organization of the colonies by division of labor (OSTER & WILSON, 1978, WILSON 1987, BOURKE & FRANKS, 1995). The adaptive skills of the honeybees deserve special mentioning compared to wasps, hornets or bumblebees in temperate regions, since they not only have to cope with highly fluctuating biotic and abiotic factors, but survive the winter as a colony and are the first to raise a new generation in spring.

The honeybee's thermoregulatory and resource delivery system has to be both stable and flexible at the same time, because thermoregulation in the hive has to be kept constant against temperature influences and with low fluctuation even if ambient temperatures rise and fall sharply. Likewise, the honey resources of the hive must be economized carefully while at the same time the blood sugar of the heating bees has to be kept constantly at a relatively high level in order to make them fulfill their thermoregulatory duty (SOTAVALTA, 1954; CRAILSHEIM, 1988; HEINRICH, 1993).

In order to comply with both requirements – stable thermoregulation and an economical distribution of the resources –, honeybees need a flexible and adaptive system relating to heating and trophallaxis in the hive.

The trophallactic activity in general guarantees a distribution of resources within the hive (NIXON & RIBBANDS, 1952) and is an important factor in making the social community work (FREE, 1959, SLEIGH, 2002). Honeybees acquire this ability within a few hours after hatching. For their first feeding contacts, they just extrude their proboscis and the antennae to feel for offered food, while after this training time (after five to six days), they use only one antenna to touch the droplet of regurgitated food which is displayed between the mandibles of the donor bee.

The antennae and the forelegs are of great importance in releasing food transfers (KORST & VELTHUIS, 1982) and in helping bees to orientate their mouthparts to one another, especially in the darkness of the hive. We found a preference in soliciting honeybees for using the right over the left antenna for touching the mouthparts of the donor bee and consequently for receiving food from the donor (**chapter two**).

General discussion

In addition, the preference of the right antenna is not a matter of individuality, because bees used the antenna alternately, whereas the right antenna was used more often than the left one (Fig. 2.1). The chemo-sensitive or gustatory properties of the antenna are responsible for a response to sugar containing liquids. A touch with sugar containing liquids triggers a PER (proboscis extension response) (KUWABARA, 1957; BITTERMANN ET AL., 1983) in the recipient which is necessary to start the actual foods transmission. The training of the antennal movement and the reduction to using only one antenna to trigger the PER simplifies the feeding contact. The probability of finding a food offer is much higher if both antennae are used independently. The preference of the right antenna might be caused by the higher sensibility for gustatory stimuli or by a general dominance of the right hemisphere over the left, since other stimuli like vision or chemotaxis are known the be lateralized to the right hemisphere in honeybees (LETZKUS ET AL., 2006, 2008).

The honeybee hive is constructed heterogenically. The brood nest is formed in the middle of the comb and on the combs that are located in the centre of the hive. The honey bearing cells are located in the periphery, separated from the brood by empty cells.

We could show that the participants of the trophallactic contacts in the brood area do not walk into one another randomly (**chapter four**). Donor bees repeatedly shuttle between honey comb, where they fill their crop with honey, and brood area, where they move towards the energy spending heating bees and offer food to hot bees. The donor bees´ orientation towards the heating bees is provided by the emitted heat itself (BASILE ET AL., 2008). Honeybees are known to discriminate temperatures of at least Δ 0.25 °C (HERAN, 1952). Therefore, they not only have the capacity to navigate to the warm brood area, where food is needed, but can also discriminate which bee in this area is actually spending energy and might be in need for refueling.

This heat dependent reward system is simple but highly effective. Since the only heat emitting source in the hive is a heat producing bee, any food offer by the donor is an offer to a bee in need, and therefore economically justifiable. Cheating is virtually impossible in this distribution system, because every hot bee is spending energy for the collective good either by heating the brood or the winter cluster, is preparing to go on a foraging flight or is heating up to defend the hive against intruders (HIMMER, 1927; ESCH, 1960; HEINRICH, 1971; SEELEY & HEINRICH, 1981A; SOUTHWICK & HELDMAIER, 1987; ONO ET AL., 1987, 1995; ESCH & GOLLER, 1991; HEINRICH, 1993; STABENTHEINER, 1996; KASTBERGER & STACHL, 2003; KEN ET AL., 2005).

General discussion

Food in general is no guarantee for sufficient heat production in honeybees, because relatively high sugar content is of crucial importance for constant high thoracic temperatures as they are necessary for the heating activity (JONGBLOED & WIERSMA, 1934; LOH & HERAN, 1970; SACKTOR, 1970; SOUTHWICK & HELDMAIER, 1987; CRAILSHEIM, 1988; ROTHE & NACHTIGALL, 1989).

Sugar solutions with low sugar content proved insufficient for raising the median thoracic temperature higher than 35 °C in worker bees without additional water (**chapter three**). Judging from these results, heating bees not only need a high quantity of food implemented by frequent food transfers, but also a high quality of food, which is reflected in the height of the sugar content of the honey which is transferred by the shuttling donor bees.

The behavioral experiments showed that donor bees shuttle frequently back and forth between the honeycomb, where they fill their crop with honey, and the brood comb where they feed several bees with relatively high thoracic temperature (BASILE ET AL., 2008).

The experiments in **chapter three** show that the ingestion of food with high sugar content and the thoracic temperature correlate positively. Consequently applied to our behavioral observations this means that if the heating bees are fed with honey which has a naturally high sugar content, they will keep their high thoracic temperature not only for behavioral (=active heating behavior) but also for physiological (=ingested sugar increases thoracic temperature) reasons.

Subject to the condition that the emitted heat from the heating bees triggers the offering behavior of the donors (**chapter five**), a self-energizing loop is created: The heat emitted by the heating bees triggers the offering of food which is high in sugar content, which in turn increases the thoracic heat production, which again makes the donor bees offer food to them etc. (Fig. 6.1). The task of the heater bees is physically exhausting but relatively simple to maintain. They have to keep their position on the brood comb or in the centre of the winter cluster, and keep their thoracic temperature elevated. After a food offering respectively a food transfer from a donor their temperature stays increased due to the high sugar content of the honey transferred from the donors.

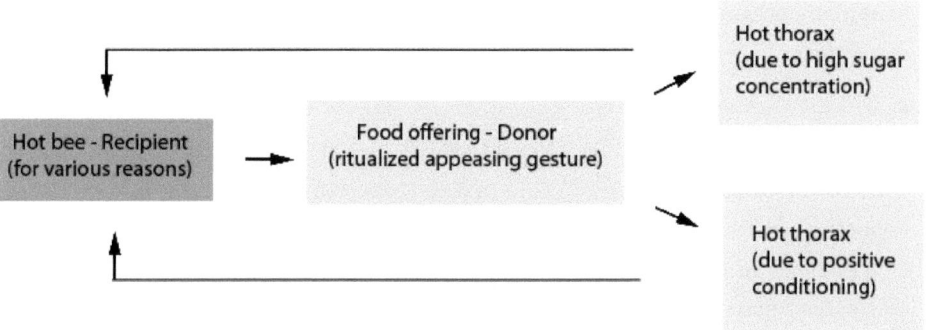

Fig 6.1 Self-energizing loop

Heat emitted by the heating bees triggers the offering of food which is high in sugar content which in turn increases the thoracic heat production, which again makes the donor bees offer food.

The task of the donors by contrast is more complicated and the behavior itself is rather difficult to comprehend.

The regurgitation of food in the presence of heat by some individuals could be explained by the behavior many animals show in the presence of aggression.

There are many different ways to show aggression in the animal kingdom. Audio, visual, and chemical cues are used to show an opponent that resistance is futile or at least that a fight or an escape reaction is about to take place. Anyway, there is a physiological condition all animals share: the body is prepared for a fight and its physical preconditions concern the muscle activity which is directly reflected in its body temperature. An aggressive individual literally "boils with rage"(DALTON, 1940).

Consequently in all aggressively acting hymenopterans an increase in thoracic temperature is observable, because the leg and wing muscles need a certain temperature to operate at full capacity. Such behavior can be observed in honeybees, when guard bees examine returning foragers or when bees attack enemies like wasps or hornets (ESCH, 1960; HEINRICH, 1971; ONO ET AL., 1987, KEN ET AL., 2005).

As mentioned before, honeybees are able to detect variations in temperatures of at least Δ 0.25 ℃. This ability gives the honeybee the capacity in a conflict situation to decide whether the opponent is in a potential "physiologically aggressive mode" or not. If the honeybee decides to avoid a conflict, it can try to appease the opponent with by food regurgitation.

Food offerings are known as appeasing gestures in many hymenopterans (Social *Vespidae* -HUNT, 1991; Porine ants - LIEBIG ET AL., 1997; Hallictine bees – KUKUK & CROZIER, 1990; Carpenter bees - VELTHUIS & GERLING, 1983; WCISLO, 2006). Even though trophallactic activities of honeybees usually have no recognizable aggressive potential, food offerings in honeybees are sometimes done for this exact reason. If for example foreign bees enter a colony or a bee gets under attack in a cage experiment, the defeated or subdominant individual often regurgitates food. Such food regurgitations of submissive individuals are generally considered as appeasing gestures in the honeybee (BUTLER & FREE, 1952; SAKAGAMI, 1954; MEYERHOFF, 1955; BREED ET AL., 1985; VAN DER BLOOM, 1991), because there is a correlation between individual worker dominance and trophallactic behavior (HILLESHEIM ET AL., 1989). The advance a dominant individual gains by receiving food from subdominant workers makes it more probable that these dominant bees can develop ovaries and become reproductive egg layers. Such positive correlation between trophallactic dominance and developing ovaries is shown by KORST and VELTHUIS (1982), LIN ET AL. (1999) and by HOOVER (2006)

Another clue for the potentially aggressive origin of the feeding contacts between heating bees and cooler donors are the reactions of the bees confronted with the heat emitting source in **chapter five**. There were mainly two different reactions depending on the individual: Some workers regurgitated food every time they were confronted with the hot carbon film resistor while others reacted rather aggressive by spreading the mandibles. Both reactions fit as answers to an aggressive thread: the food offering as an appeasing and submissive gesture and the aggressive spreading of the mandibles as a dominant gesture. Our experimental setup allowed only little reaction, since the bees were immobilized – except from their antennae and mouthparts— and they were unable to avoid the heat source. In our experiment with the dead bees which were equipped with a carbon film resistor and heated one at a time, some bees of our experiment chose to move towards the hot bees and repeatedly offered food to them voluntarily. The individual differences between the bees´ reactions to a heat emitting source might have many different reasons: We picked the bees randomly from the brood comb. Their reaction could be linked to their JH status, which is known to influence aggressive behavior (PEARCE ET AL., 2001). Another factor influencing their behavior towards a heat emitting source could be their age. Even though bees on the brood comb are considered as relatively young bees occupied with different tasks of brood care, there is no guarantee that the bees in the experiment were of the same age. The genetic variance between the bees in the experiment might be a strong factor for different behavior as well. Since guarding as a rather aggressive and dominant behavior and the exchange of food as a submissive behavior is known to be influenced by the genetic background of a worker (ROBINSON & PAGE, 1988; MORITZ & HILLESHEIM, 1985; HILLESHEIM, 1989), the reaction to heat emitting objects might be influenced by the genetic background as well.

The behavior of donors on the brood comb which voluntarily move towards the hot bees and offer food to them in **chapter four** is comparable to the behavior some bees showed in the experiments in **chapter five**. However, the shuttling behavior of donors between brood comb and honeycomb in **chapter four** still requires an explanation.

HOGEWEG and HESPER (1983, 1985) showed with their "Mirror" model that minimal conditions are needed for the formation of the two types of workers in bumblebees. The artificial bumblebees´ behavior is triggered by what they encounter. When an adult meets another, a dominance interaction takes place, the outcome of which (dominant or submissive behavior) is self-reinforcing. This model automatically generates two stable classes, those of "commons" (low-ranking workers), which are busy with foraging in the periphery of the nest and "elites" (high ranking workers), which care for the brood occupy the centre of the nest.

Provided that the donors and the heating bees in **chapter four** and **five** show signs of a dominance hierarchy where the donors display submissive behavior by offering food and recipients display dominant behavior by heating and getting fed, there are obvious parallels between the bumblebee model and the observation of the honeybee behavior on the brood comb:

The "aggressive" heating bees occupy the centre of the nest, interact more often with the queen and one another (due to the cramped premises on the brood comb) and a dominance interaction with a submissive worker (donor) ends with a gain in food for the "aggressive" heating bee.

The "submissive" donor bees occupy the periphery more often than the "aggressive" bees and are busy not exactly foraging but collecting and distributing the resources.

If HAMILTON´S (1971) selfish-herd theory and the centripetal instinct applies to the social interaction of the honeybee as well, the dominant individuals should be found in the centre of the "herd". In our case, the "aggressive" or dominant hot bees do occupy the very centre of the nest, which is the brood comb respectively the centre of the winter cluster, which is consistent with the distribution of the two "classes" of bees and their behavior in the observations of **chapter four** and **five**.

Social dominance and the relationship between dominance and aggression are considered of fundamental social importance (GARTLAN, 1968; FRANCIS, 1988). There are two opposing views. On one hand a higher rank is believed to offer optimal access to resources, and therefore individuals should seize every opportunity to increase in rank (POPP & DEVORE, 1979). In the present case, the dominant heater bees do not have direct access to food, because they are separated from the honey comb, but their heating activity makes sure that donor bees offer food to them repeatedly which could be characterized as "access" to food of another type. As long as they keep the elevated thoracic temperature, the supply with food is guaranteed and they keep their rank and position in the middle of the nest.

On the other hand, the function of a dominance hierarchy is thought to reduce costs associated with aggression and therefore, individuals should avoid conflict as soon as relationships are clear (POPP & DEVORE, 1979). This present case of dominance interaction in the honeybee has no open conflict; therefore costs for the individual in terms of injuries are eliminated. By contrast, open conflicts are known in queenless honeybee colonies, when workers are competing for becoming a "pseudo-queen" (SAKAGAMI, 1954; SEELEY, 1985; RATNIEKS, 1988).

Instead, the heat triggered feeding system on the honeybees´ brood comb as described in **chapter four** does reduce costs very efficiently. The described system is time-saving, in avoiding the begging and potential refusing of regurgitation, and energy-saving at the same time. The abundance of recipients on the brood comb increases brood rearing efficiency (SCHOLZE, 1964; SOUTHWICK & HELDMAIER, 1987; CRAILSHEIM, 1988) and improves the efficiency of insulation against heat loss (SACKTOR, 1970, KRONENBERG & HELLER, 1982; SOUTHWICK & HELDMAIER, 1987).

A similar activity to that of donors refueling heat-emitting bees (**chapter four** and **five**) has been described for foragers by BRANDSTETTER ET AL. (1988). The feeding of foragers in the hive might work on the same principle, because foragers usually attain elevated thoracic temperatures in the hive before they leave for their flight.

General discussion

The connection between heat emission, which is a sign for aggression, and food offering, which is an appeasing gesture, might be the main steps in the process of signal evolution. If a behavior becomes stereotyped and changes function it is referred to as ritualized (TINBERGEN, 1952). It is generally recognized that ritualization has played a major role in the evolution of communication in social insects (HÖLLDOBLER, 1984; WILSON, 1985B) especially in the honeybees´ dance communication. Only recently, RITTSCHOF and SEELEY (2008) presented a case of ritualization in the honeybee´s buzz-run.

The process of signal evolution is described as a procedure, in which the establishment of an association between the particular condition of the sender and the production of the cue by the sender is the first step (OTTE, 1974). In our case, the condition of heating and the cue "heat" are easily associated with one another, because the cue is an inevitable byproduct of heating.

The benefit of this ritualized appeasing gesture is evident, since the offering behavior is target-oriented towards energy-consuming hot bees (Fig. 6.2). Every hot individual contributes to the welfare of the hive, and in addition, the offering is only accepted if the threshold for sugar is low enough in the recipient i.e. if the recipient is in need.

Comparing this elaborate energy distributing system with related groups, like bumblebees, or social wasps, leads to the conclusion that there is no equivalent system. In other hymenopteran societies brood heating is either not regulated strictly, or there is no trophallaxis between the adult members of a colony in non-apine hymenopterans. Apart from that, a winter cluster is a phenomenon solely formed in honeybees. Only the honeybee has evolved to a point that enables the colony perennial survival and provides the opportunity to produce the next generation of workers as soon as spring provides slightly warmer ambient temperatures and enough pollen.

This system of providing the heat-emitting bees by using their heat as a cue and a formerly appeasing gesture as ritualized response, contributes to the highly economical resource management that is in line with the ecological requirements in the colony and simultaneously with the physiological conditions of the individual. Our finding of the perpetual heating by an instant refill system gives insight to a complex, but highly economical system which regulates itself und fulfills the challenging task of keeping the brood nest or the core of the winter cluster constantly warm. Only this system makes the honeybees independent from ambient temperature, even if sudden cold spells exacerbate

General discussion

the process, and it is an ability only honeybees have accomplished among the social insects.

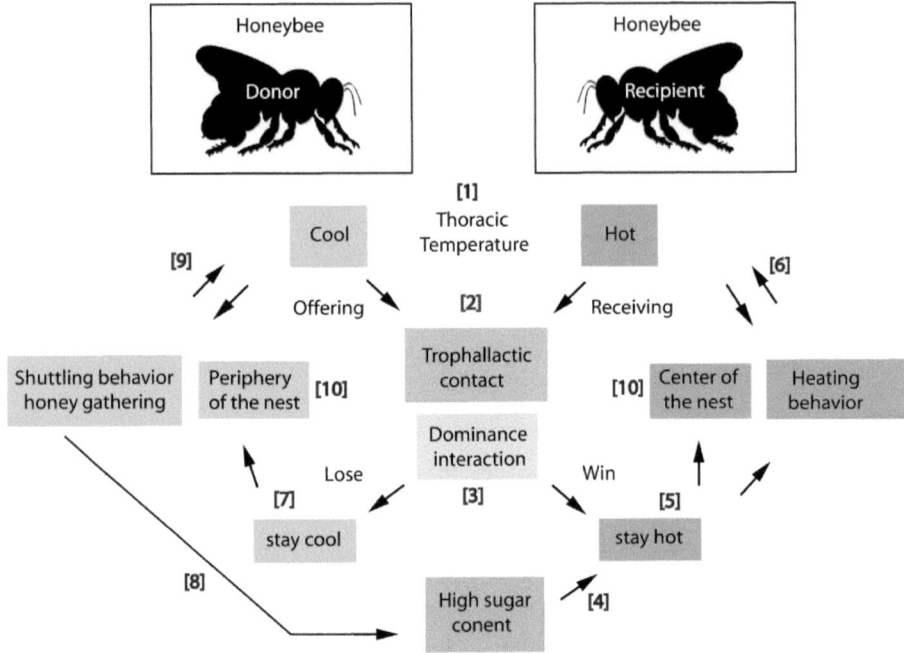

Fig. 6.2 Summarizing diagram

[1] The thoracic temperatures of donors and recipients of a trophallactic contact differ according to their role.
[2] The recipients have a higher thoracic temperature and the donors have a cooler thoracic temperature.
[3] The regurgitation of food is described as a submissive appeasing gesture in many hymenopterans. A hot thorax is a side-effect of aggressive behavior. Therefore, a trophallactic contact can also be seen as a dominance interaction.
[4] High sugar content food causes higher thoracic temperatures.
[5] and [6] The hot thorax of the „winning" recipient guarantees constant „refill" for the heating bee (and it will continue its heating behavior).
[7] The donor bees which are cooler then the recipient bees stay cool and shuttle back to the periphery of the nest (honeycomb) where they reload on honey.
[8] After reloading, they shuttle back to the brood comb, where they feed hot bees with the high sugar content honey which again makes them stay hot [4].
[9] The shuttling behavior and the constant „losing" of food (losing the dominance interaction) makes them repeat their offering behavior.
[10] The shuttling to the honeycomb makes the donors spend more time in the periphery of the nest. The heating bees stay in the center of the nest, which is why they spend more time in the center then the donors.

7. Summary

Like many other social insect societies, honeybees collectively share the resources they gather by feeding each other. These feeding contacts, known as trophallaxis, are regarded as the fundamental basis for social behavior in honeybees and other social insects for assuring the survival of the individual and the welfare of the group. In honeybees, where most of the trophallactic contacts are formed in the total darkness of the hive, the antennae play a decisive role in initiation and maintenance of the feeding contact, because they are sensitive to gustatory stimuli. The sequences of behaviors performed by the receiver bees at the beginning of a feeding contact includes the contact of one antenna with the mouthparts of a donor bee where the regurgitated food is located. The antennal motor action is characterized by behavioral asymmetry, which is novel among communicative motor actions in invertebrates. This preference of right over left antenna is without exception even after removal of the antennal flagellum. This case of laterality in basic social interaction might have its reason in the gustatory asymmetry in the antennae, because the right antenna turns out to be significantly more sensitive to stimulation with sugar water of various concentrations than the left one.

Trophallactic contacts which guarantee a constant access to food for every individual in the hive are vitally important to the honeybee society, because honeybees are heterothermic insects which actively regulate their thoracic temperature. Even though the individual can regulate its body temperature, its heating performance is strictly limited by the amount of sugar ingested. The reason for this is that honeybees use mostly the glucose in their hemolymph as the energy substrate for muscular activity, and the heat producing flight muscles are among the metabolically most active tissues known. The fuel for their activity is honey; processed nectar with a sugar content of ~80% stored in the honeycomb. The results show that the sugar content of the ingested food correlates positively with the thoracic temperature of the honeybees even if they are caged and show no actual heating-related behavior as in brood warming or heating in the centre of the winter cluster.

Honeybees actively regulate their brood temperature by heating to keep the temperature between 33 °C to 36 °C if ambient temperatures are lower. Heating rapidly depletes the worker's internal energy; therefore the heating performance is limited by the honey that is ingested before the heating process. This study focused on the behavior and the thoracic

Summary

temperature of the participants in trophallactic food exchanges on the brood comb. The brood area is the centre of heating activity in the hive, and therefore the region of highest energy demand. The results show that the recipients in a trophallactic food exchange have a higher thoracic temperature during feeding contacts than donors, and after the feeding contact the former engage in brood heating more often. The donor bees have lower thoracic temperature and shuttle constantly between honey stores and the brood comb, where they transfer the stored honey to heating bees.

In addition, the results show a heat-triggered mechanism that enables donor and recipient to accomplish trophallactic contacts without delay in the total darkness of the hive in the brood area as the most energy consuming part of the hive.

Providing heat-emitting workers with small doses of high performance fuel contributes to an economic distribution of resources consistent with the physiological conditions of the bees and the ecological requirements of the hive, resulting in a highly economical resource management system which might be one of the factors favouring the evolution of perennial bee colonies in temperate regions.

The conclusion of these findings suggests a resource management strategy that has evolved from submissive placation behavior as it is seen in honeybees, bumblebees and other hymenopterans. The heat-triggered feedback mechanism behind the resource management of the honeybee´s thermoregulatory behavior reveals a new aspect of the division of labor and a new aspect of communication, and sheds new light on sociality in honeybees.

8. Index of Figures

Fig. 1.1 Coefficient of relatedness in diploid organism p. 9

Fig. 1.2 Coefficient of relatedness with haplo-diploid sex determination p. 11

Fig. 1.3 European subspecies of *Apis mellifera* and their distribution p. 15

Fig. 1.4 Hormone levels in the honeybee worker in relation with age and labor p. 21

Fig. 1.5 Trophallactic contact in *A. mellifera* p. 31

Fig. 2.1 Feeding activities of recipient with both antennae intact p. 50

Fig. 2.2 Feeding activities of recipient with right antenna amputated p. 51

Fig. 2.3 Feeding activities of recipient with left antenna amputated p. 52

Fig. 2.4 Test for individual preference of one antenna p. 53

Fig. 2.5 Gustatory response scores p. 54

Fig. 3.1 Thoracic temperatures in worker bees fed with different sugar solutions without additional water p. 75

Fig. 3.2 Maximal thoracic temperatures in worker bees fed with different sugar solutions without additional water p. 75

Fig 3.3 Thoracic temperatures in worker bees fed with different sugar solutions with additional water p. 76

Fig. 3.4 Maximal thoracic temperatures in worker bees fed with different sugar solutions with additional water p. 76

Fig. 3.5 Possible influence of water, sugar and water produced by energy metabolism on the thoracic temperature of caged honeybees p. 77

Index of Figures

Fig. 4.1 Map depicting the position of brood comb and honeycomb in a two frame observation hive   p. 100

Fig. 4.2 Coherence between thoracic temperature and role in trophallactic contact   p. 101

Fig. 4.3 Number of times (frequency) donors and recipients bridged the distance between brood comb and honeycomb   p.102

Fig. 4.4 Behavioral patterns of donors and recipients   p. 103

Fig. 4.5 Percentages of donors and recipients ever showing feeding or heating behaviour   p. 104

Fig. 4.6 Feeding contact frequency and duration   p. 105

Fig. 5.1 Schematized comb with four different areas and boxplot with trophallactic indices   p. 123

Fig. 5.2 Shift of body axis of donor and recipient before and during trophallaxis   p. 124

Fig. 5.3 Schematized comb with winter cluster and boxplot of warm and cold frame trophallactic indices   p. 125

Fig. 5.4 Schematized comb with winter cluster and boxplot of warm, cold and warmed-up frame trophallactic and offering indices   p. 126

Fig. 5.5 Offering and begging behavior towards hot and cold dead bees   p.127

Fig. 5.6 Frequency of behavior of restrained bees towards a warm and cold carbon film resistor   p. 128

Fig. 5.7 Frequency of behavior towards carbon film and high power wound resistor   p. 128

Fig. 5.8 Trophallactic interaction of donor and recipient on the brood comb   p. 129

Index of Figures

Fig. 6.1 Self-energizing loop   p. 137

Fig. 6.2 Summarizing diagram   p. 142

Suppl. Fig. 1 Participation of left and right antenna and the respective stump   p. 55

Suppl. Fig. 2 Measured temperature value and the deviation by the foil   p. 78

9. Index of Tables

Tab. 1.1 Sensilla present on the honeybees´ antennae and stimuli they are receptive to   p. 36

Tab. 2.1 Antennal uses in trophallactic activity   p. 56

Tab. 3.1 Thoracic temperatures (all bees per picture) of worker bee groups fed with different sugar concentrations without additional water   p. 79

Tab. 3.2 Thoracic temperatures (only hottest bee per picture) of worker bee groups fed with different sugar concentrations without additional water   p. 80

Tab. 3.3 Thoracic temperatures (only hottest bee per picture) of worker bee groups fed with different sugar concentrations and additional water   p. 81

Tab. 3.4 Thoracic temperatures (only hottest bee per picture) of worker bee groups fed with different sugar concentrations and additional water   p. 81

Tab. 5.1 Trophallactic indices on different areas of the comb   p. 130

Tab. 5.2 Trophallactic indices on different areas of the winter cluster   p. 131

Tab. 5.3 Offering indices on different areas of the winter cluster   p. 132

Tab. 5.4 List of all observed behaviors towards hot and cold carbon film resistors and p-values   p. 133

Tab. 5.5 List of all observed behaviors towards carbon film and high power resistors and p-values   p. 133

Suppl. Tab. 1 Antennal use in trophallactic activity   p. 56

Suppl. Tab. 2 Differences between measured temperature with and without foil, influence (Δ °C) produced by the foil (error) which has to be added or subtracted from the measured temperature   p. 82

Suppl. Tab. 3 Sugar concentrations used in chapter 3 in percent and mol   p. 83

Index of Tables

Suppl. Tab. 4 Multiple comparisons of all groups: p-levels for groups of worker bees without additional water   p. 84

Suppl. Tab. 5 Multiple comparisons for all groups: p-levels for groups of worker bees without additional water (only hottest bee per group picture)   p. 85

Suppl. Tab. 6 Multiple comparisons for all groups: p-levels for groups of worker bees with additional water   p. 86

Suppl. Tab. 7 Multiple comparisons for all groups: p-levels for groups of worker bees with additional water (only hottest bee per group and picture)   p. 86

10. Index of Abbreviations

Fig. = figure

Tab. = table

Suppl. = supplementary

c = cost for the individual

b = benefit for the kin

r = relatedness

♀ = female

♂ = male

% = percent

JH = juvenile hormone

p/mol = picomol

µl = micro liter

Zn = zinc

mm = millimeter

g = gram

min = minute

°C = degree Celsius

i.e. = id est

e.g. = exemplī grātia

a.m. = ante meridiem

p.m. = post meridiem

Index of Abbreviations

± = numerical value of a quantity together with its tolerance

PER = proboscis extension response

h = hour

x = multiplication sign

X^2 = chi-square distribution

n = sample size

df = degrees of freedom

p = p-value (probability of obtaining a result at least as extreme as the one actually observed)

Z = Z-score (Standard score)

SD = Standard deviation

SE = Standard error

n.a. = not applicable

n.s. = not significant

CO_2 = carbon dioxide

ATP = Adenosine triphosphate

cm = centimeter

© = copyright

™ = trade mark

U.S. = United States

Inc. = Incorporation

Δ = delta (interval of possible values for a given quantity)

R = Pearson product-moment correlation coefficient

Q = quartile

Index of Abbreviations

Min = minimum

Max = maximum

mol = mole

V = Volt

A = Ampere

* = $p<0.05$

** = $p<0.01$

*** = $p<0.001$

etc. = et cetera

11. References

ABOU-SEIF M.A.M., MAIER V., FUCHS J., MEZGER M., PFEIFFER E.F., BOUNIAS M. (1993) Fluctuations of carbohydrates in haemolymph of honeybee (*Apis mellifica*) after fasting, feeding and stress. Hormone and Metabolic Research 25, 4-8.

ADAMS J., ROTHMAN E.D., KERR W.E., PAULINO Z.L. (1977) Estimation of the number of sex alleles and queen matings from diploid male frequencies in a population of *Apis mellifica*. Genetics 86, 583-596.

ADES C., RAMIRES N. (2002) Asymmetry of leg use during prey handling in the spider *Scytodes globula* (Scytodidae). Journal of Insect Behavior 15, 563-570.

ALEXANDER R.D. (1961) Aggressiveness, territoriality, and sexual behaviour in field crickets (Orthoptera: Gyllidae). Behaviour 17, 130-223.

ALLEN M.D. (1955) Observations of honeybees attending their queen. British Journal of Animal Behaviour 3, 66-69.

ALLEN M.D. (1960) The *honeybee queen and her* attendants. Animal Behaviour 8, 201-208.

ALTMANN G. (1956) Die Regulation des Wasserhaushaltes der Honigbiene. Insectes Sociaux 3, 33-40.

ALTMANN G., GONTARSKI H. (1963) Über den Wasserhaushalt der Winterbienen. Symposia Genetica et Biologica Italica 12, 308-328.

AMDAM G.V., OMHOLT S.W. (2002) The regulatory anatomy of honeybee lifespan. Journal of Theoretical Biology 216, 209-228.

AMDAM G.V, AASE A.L.T.O., SEEHUUS S., FONDRK M.K., NORBERG K., HARTFELDER K. (2005) Social reversal of immunosenescence in honey bee workers. Experimental Geronotlogy 40, 939-947.

ARNOLD G., QUENET B., CORNUET J.M., MASSON C., DE SCHEPPER B., ESTOUP A., GASQUI P. (1996) Kin recognition in honey bees. Nature 379, 498.

BAKER H.G., BAKER I. (1983) *A Brief Historical Review of the Chemistry of Floral Nectar*. In: Bentley B., Elias T. (Eds.) The Biology of Nectaries. Columbia University Press, New York, USA, pp. 126-152.

References

BASILE R., PIRK C.W.W, TAUTZ J. (2008) Trophallactic activities in the honeybee brood nest - Heaters get supplied with high performance fuel. Zoology 111, 433-441.

BEETSMA J. (1979) The process of queen-worker differentiation in the honeybee. Bee World 660, 24-39.

BERTSCH A. (1984) Foraging in male bumblebees (*Bombus lucorum* L.): maximizing energy or minimizing water load? Oecologia 62, 325-336.

BEUTLER R. (1950) Zeit und Raum im Leben der Sammelbiene. Naturwissenschaften 37, 102-105.

BEYE M., HASSELMANN M., FONDRK M.K., PAGE R., OMHOLTS S.W. (2003) The gene csd ist he primary signal for sexual development in the honeybee and encodes an SR-type protein. Cell 114, 419-429.

BITTERMANN M.E., MENZEL R., FIETZ A., SCHÄFER S. (1983) Classical conditioning of proboscis extension in honeybees (*Apis mellifera*), Journal of Comparative Psychology 97, 107-119.

BLATT J., ROCES F. (2001) Haemolymph sugar levels in foraging honeybees (*Apis mellifera carnica*): dependence on metabolic rate and in vivo measurement of maximal rates of trehalose synthesis. Journal of Experimental Biology 204, 2709-2716.

VAN DER BLOOM J. (1991) Social regulation of egg-laying by queenless honeybee workers (*Apis mellifera* L.). Behavioral Ecology and Sociobiology 29, 431-436.

BLOUNT B.G. (1990) Issues in bonobo (*Pan paniscus*) sexual behavior. American Anthropologist 92, 702-14.

BOCH R. (1957) Rassenmässige Unterschiede bei den Tänzen der Honigbiene (*Apis mellifica* L.) Zur vergleichenden Physiologie. 40, 289-320.

BODENHEIMER F.S. (1937) Studies in animal populations. II. Seasonal population-trends of the honey-bee. The Quarterly Review of Biology 12, 406-425.

BOURKE A.F.G., FRANKS N.R. (1995) *Social evolution in ants*. Monographs in Behavioral Ecology. Princeton University Press, Princeton, New Jersey. USA.

BRANDSTETTER M., CRAILSHEIM K., HERAN H. (1988) Provisioning of food in the honeybee before foraging. Biona Report 6, 129-148.

References

BREED M.D., BUTLER L., STILLER T.M. (1985) Kin recognition by worker honeybees in genetically mixed groups. Proceedings of the National Academy of Sciences 82, 3058-3061.

BUJOK B., KLEINHENZ M., FUCHS S., TAUTZ J. (2002) Hot spots in the bee hive. Naturwissenschaften 89, 299-301.

BUJOK B. (2005) *Thermoregulation im Brutbereich der Honigbiene Apis mellifera carnica.* PhD Dissertation. University Würzburg.

BURGESS E. P. J., MALONE L. A., CHRISTELLER J. T. (1996) Effects of two proteinase inhibitors on the digestive enzyme. Journal of Insect Physioogy 42, 823-828.

BUTLER C.G. (1940) The choice of drinking water by the honeybee. Journal of Experimental Biology 17, 253-261.

BUTLER C.G., FREE J.B. (1952) The behaviour of worker honeybees at the hive entrance. Behaviour 4, 262-292.

BUTLER C.G. (1954) The method and importance of the recognition by a colony of honeybees of the presence of its queen. Transactions of the Royal Entomological Society London 105, 11-29.

BUTLER C.G. (1957) The process of queen supersedure in colonies of honeybees *(Apis mellifera* Linn.). Insectes Sociaux 4, 213-223.

BUTLER C.G. (1960) Sex determination and caste differentiation in the honeybee (*Apis mellifera*). Memoir 7, Society for Endocrinology, 3-8.

CAMAZINE S. (1991) Self-organizing pattern formation on the combs of honey bee colonies. Behavioral Ecology and Sociobiology 28, 61-76.

CANDY D. J., BECKER A., WEGENER G. (1997) Coordination and integration of metabolism in insect flight. Comparative Biochemistry and Physiology 117B, 497-512.

CORONA M., VELARDE R.A., REMOLINA S., MORAN-LAUTER A., WANG Y., HUGHES K.A., ROBINSON G. (2007) Vitellogenin, juvenile hormone, insulin signaling, and queen honey bee longevity. Proceedings of the National Academy of Science 11, 7128-7133.

References

CRAILSHEIM K. (1988) Intestinal transport of glucose solution during honeybee. Biona Report 6, 119-128.

CRAILSHEIM K. (1990) The protein balance of the honey bee worker. Apidologie 21, 417-429.

CRAILSHEIM K. (1992) The flow of jelly within a honeybee colony. Journal of Comparative Physiology B: Biochemical, Systemic, and Environmental Physiology 162, 681-689.

CRAILSHEIM K., LEONHARD B. (1997) Amino acids in honeybee worker haemolymph. Amino Acids 13, 141-153.

CRAILSHEIM K. (1998) Trophallactic interactions in the adult honeybee (*Apis mellifera* L.) Apidologie 29, 97-112.

CRANE E. (1996) The removal of water from honey. Bee World 77, 120–129.

CRANE E. (1999) *The World History of Beekeeping and Honey Hunting*, Routledge Chapman & Hall, New York, New York, USA.

CREMONEZ T.A., DE JONG D., BITONDI M.M. (1998) Quantification of hemolymph proteins as a fast method for testing protein diets for honeybees. Journal of economic entomology 91, 1284-1289.

CROSLAND M.W.J. (1990) Variation in ant aggression and kin discrimination ability within and between colonies. Journal of Insect Behaviour 3, 359-379.

CROZIER R.H. (1977) Evolutionary genetics of the hymenoptera. Annual Review of Entomology 22, 263-288.

DALTON D.J. (1940) The effect of maintenance of normal body temperature during the alarm reaction. Anatomical Records 78, 110-111.

DAWKINS R. (1976) *The Selfish Gene*. A Paladin Book, London: Granada, United Kingdome (published 1978).

DAWKINS R. (1982) The Extended Phenotype. Oxford University press, W.H. Freeman and Company, United Kingdome.

DEGROOT A.P. (1952) Amino acid requirements for growth of the honeybee (*Apis mellifica* L.). Experentia 8, 192-194.

References

DETROY B.F., WHITEFOOT L.O., MOELLER F.E. (1981) Food requirements of caged honey bee (Apis mellifera). Apidologie 12, 113-124.

DORNHAUS A., KLÜGL F., PUPPE F., TAUTZ J. (1998) Task Selection in Honeybees - Experiments Using Multi-Agent Simulation. Verlag Harry Deutsch AG, Bochum, Germany.

DYER F. C., SEELEY T.D. (1987) Interspecific comparisons of endothermy in honey-bees (APIS): Deviations from the Expected Size-Related Patterns. Journal of Experimental Biology 127, 1-26.

ENGELS W., FAHRENHORST H. (1974) Alters- und kastenspezifische Veränderungen der Haemolymph-Protein-Spektren bei Apis mellificia. Rouxs Archives of Developmental Biology 174, 285-296.

ENGELS W., KAATZ H-H., ZILLIKENS A., SIMÕES Z.L.P., TRUBE A., BRAUN R., DITTRICH F. (1990) Honey bee reproduction: vitellogenin and caste-specific regulation of fertility. Advances in Invertebrate Reproduction 5, 495-502.

ESCH H.E. (1960) Über die Körpertemperaturen und den Wärmehaushalt von. Apis mellifica. Zeitschrift für Vergleichende Physiologie 43, 305-335.

ESCH H.E., GOLLER F. (1991) Neural control of fibrillar muscles in bees during shivering and flight. Journal of Experimental Biology 159, 419-431.

ESSLEN J., KAISSLING K.E. (1976) Zahl und Verteilung antennaler Sensillen bei der Honigbiene (Apis mellifera L.). Zoomorphology 83, 227-251.

ESTOUP A., SOLIGNAC M., CORNUET J.M. (1994) Precise assessment of the number of patrilines and of genetic relatedness in honey bee colonies. Proceedings of the Royal Society of London B 258, 1-7.

FAHRBACH S.E., ROBINSON G.E. (1996) Juvenile hormone. Behavioral maturation, and brain structure in the honey bee. Developmental Neuroscience 18, 102-114.

FALCHUK K.H. (1998) The Molecular Basis for the Role of Zinc in Developmental Biology, Molecular and Cellular Biochemistry 188, 41-48.

FARINA W.M., NÚÑEZ J. A. (1991) Trophallaxis in the honeybee, Apis mellifera (L.), as related to the profitability of food sources. Animal Behaviour 42, 389-394.

References

FARINA W.M., NÚÑEZ J.A. (1995) Trophallaxis in *Apis mellifera*. Effects of sugar concentration and crop load on food distribution. Journal of Apicultural Research 34, 93-96.

FARINA W.M, WAINSELBOIM A.J. (2001A) Thermographic recordings show that honeybees may receive nectar from foragers even during short trophallactic contacts. Insects Socieaux 48, 360-362.

FARINA W.M, WAINSELBOIM A.J. (2001B) Changes in the thoracic temperature of honeybees while receiving nectar from foragers collecting at different reward rates. Journal of Experimental Biology 204, 1653-1658.

FARINA W.M, WAINSELBOIM A.J. (2005) Trophallaxis within the dancing context: a behavioral and thermographic analysis in honeybees (*Apis mellifera*). Apidologie 36, 43-47.

FLURI P., LÜSCHER M., WILLE H., GERIG L. (1982) Changes in weight of the pharyngeal gland and

haemolymph titers of juvenile hormone, protein and vitellogenin in worker honey bees. Journal of Insect Physiology 28, 61-68.

FLURI P., BOGDANOVS S. (1987) *Age Dependence of Fat Body Protein in Summer and Winter Bees* (*Apis mellifera*). In: Eder K., Rembold H. (Eds.), Chemistry and Biology of Social Insects, Verlag J. Peperny, München, Germany, pp. 170-171.

FRANCIS R.C. (1988) On the relationship between aggression and social dominance. Ethology, 78, 223-237.

FREE J.B. (1956) A study of the stimuli which release the food begging and offering responses of worker honeybees. British Journal of Animal Behaviour 4, 94-101.

FREE J.B. (1957) The transmission of food between worker honeybees. British Journal of Animal Behaviour, 41-47.

FREE J.B., BUTLER C.G. (1958) The size of apertures through which worker honeybees will feed one another. Bee World 39, 40-42.

References

FREE J.B., SPENCER-BOOTH Y.H. (1958) Observations on the Temperature Regulation and Food Consumption of Honeybees (*Apis Mellifera*). Journal of Experimental Biology 35, 930-937.

FREE J.B. (1959) The transfer of food between the adult members of a honeybee community. Bee World 40, 193-201.

FREE J.B., SPENCER-BOOTH Y.H. (1959) The longevity of worker honey bees. Proceedings of the Royal Entomological Society London (A) 34, 141-150.

FREE J.B., SPENCER-BOOTH Y.H. (1960) Chill-coma and cold death temperatures of *Apis mellifica*. Entomologia Experimentalis et Applicata 3, 222-230.

FREE J.B. (1961) Hypopharyngeal gland development and division of labour in honeybee (*Apis mellifera* L.) colonies. Proceedings of the Royal Entomological Society London (A) 36, 5-8.

FREE J.B. (1965) The allocation of duties among worker honeybees. Symposia of the Zoological Society of London 14, 39-59.

FREE J.B., FERGUSON A.W., SIMPKINS, J.R. (1992) The behaviour of queen honeybees and their attendants. Physiological Entomology 17, 43-55.

VON FRISCH K., RÖSCH G.A. (1926) Neue Versuche über die Bedeutung von Duftorgan und Pollenduft für Verständigung im Bienenvolk. Zeitschrift für Vergleichende Physiologie 4, 1-21.

VON FRISCH K. (1965) *Tanzsprache und Orientierung der Bienen*. Springer, Berlin, Germany.

VON FRISCH K. (1967) *The Dance Language and Orientation of Bees*, Harvard University Press, Cambridge, Massachusetts, USA.

FUCHS S., MORITZ R.F.A. (1999) Evolution of extreme polyandry in the honeybee *Apis mellifera* L. Behavioral Ecology and Sociobiology 45, 269-275.

FUKUDA H. (1983) The Relationship between work efficiency and population size in a honeybee colony. Researches on Population Ecology 25, 249-263.

GARTLAN J.S. (1968) Structure and function in primate society. Folia Primatologica 8, 89-120.

References

GOULD J.L., KIRSCHVINK J.L., DEFFEYES K.S. (1978) Bees have magnetic remanence. Science, 210, 1026-1028.

GOULD J.L., KIRSCHVINK J.L., DEFFEYES K.S., BRINES M.L. (1980) Orientation of demagnetized bees. Journal of Experimental Biology 86, 1-8.

GOULD J.L. (1980) The case for magnetic-field sensitivity in birds and bees. American Scientist 68, 256-267.

GOULD J.L. (1982) Why do honey bees have dialects? Behavioral Ecology and Sociobiology 10, 53-56.

GROH C., TAUTZ J., RÖSSLER, W. (2004) Synaptic organization in the adult honey bee brain is influenced by brood-temperature control during pupal development. Proceedings of the National Academy of Science 101, 4268-4273.

HAMILTON W.D. (1964) The genetical theory of social behaviour, Journal of Theoretical Biology 7, 1-16 and 17-32.

HAMILTON W.D. (1971) Geometry for the selfish herd. Journal of Theoretical Biology 31, 295-311.

HAMILTON W.D. (1972) Altruism and related phenomena, mainly in social insects. Annual Review of Ecology Evolution and Systematics 3, 193-232.

HANSER G., REMBOLD H. (1960) Über den Weiselzellenfuttersaft der Honigbiene. IV. Jahreszeitliche Veränderungen im Biopteringehalt des Arbeiterinnenfuttersaftes. Hoppe-Seyler's Zeitschrift für Physikalische Chemie 319, 200-205.

HANSER G., REMBOLD H. (1964) Analytische und histologische Untersuchungen der Kopf- und Thoraxdrüsen bei der Honigbiene *Apis mellifera*. Zeitschrift für Naturforschung B 19, 938-943.

HARRISON J. M. (1986) Caste-specific changes in honeybee flight capacity. Physiological Zoology 59, 175-187.

HAUNERLAND N.H., SHIRK P.D. (1995) Regional and functional differentiation in the insect fat body. Annual Review of Entomology 40, 121-145.

HAUPT S.S. (2004) Antennal sucrose perception in the honey bee (Apis mellifera L.): behaviour and electrophysiology. Journal of Comparative Physiology A 190, 735-745.

References

HAUPT S.S., Klemt W. (2005) Habituation and dishabituation of exploratory and appetitive responses in the honey bee (Apis mellifera). Behavioural Brain Research 165, 12-17.

HAUPT S.S. (2007) Central gustatory projections and side-specificity of operant antennal muscle conditioning in the honeybee. Journal of Comparative Physiology A 193, 523-535.

HAYDAK M.H. (1957) Changes with age in the appearance of some internal organs of the honey bee. Bee World 38, 197-203.

HAYDAK M.H. (1963) Age of nurse bees and brood rearing. Journal of Apicultural Research 2, 101-103.

HAYDAK M.H. (1970) Honey Bee Nutrition. Annual Review of Entomology 15, 143-156.

HEINRICH B. (1971) Thermoregulation of African and European honeybees during foraging, attack, and hive exits and returns. Journal of Experimental Biology 80, 217–229.

HEINRICH B. (1981A) *Insect Thermoregulation.* John Wiley & Sons. New York, New York, USA.

HEINRICH B. (1981B) Energetics of honeybee swarm thermoregulation. Science 212, 565–566.

HEINRICH B. (1993) *The hot-blooded insects: strategies and mechanisms of thermoregulation.* Harvard University Press, Cambridge, Massachusetts, USA.

HELLMICH R.L., ROTHENBUHLER W.C. (1986) Relationship between different amounts of brood and the collection and use of pollen by the honey bee (Apis mellifera). Apidologie **17**, 13-20.

HEMELRIJK C.K. (2002) Understanding of social behaviour with the help of complexity science. Ethology 108, 655-671.

HEPBURN H. R., ARMSTRONG E., KURSTJENS S. (1983) The ductility of native beeswax is optimally related to honeybee colony temperature. South Africa Journal Science 79, 416-417.

HEPBURN H. R. (1986) *Honeybees and Wax.* Springer Verlag, Berlin, Germany.

References

HEPBURN H.R., RADLOFF S. (1998) Population structure and morphometric variance of the *Apis mellifera scutellata* group of honeybees in Africa. Genetics and Molecular Biology 23, 305-316.

HERAN H. (1952) Untersuchungen über den Temperatursinn der Honigbiene (*Apis mellifera*) unter besonderer Berücksichtigung der Wahrnehmung strahlender Wärme. Journal of Comparative Physiology A 34, 179-206.

HEUTS B.A., BRUNT T. (2005) Behavioral left-right asymmetry extends to arthropods. Behavioral Brain Sciences 28, 601-602.

HILLESHEIM E., KOENIGER N., MORITZ R.F.A. (1989) Colony performance in honeybees (*Apis mellifera capensis* Esch.) depends on the proportion of subordinate and dominant workers. Behavioral Ecology and Sociobiology 24, 291-296.

HIMMER A. (1925) Körpertemperaturmessungen an Bienen und anderen Insekten. Erlanger Jahrbuch für Bienenkunde 3, 44-115.

HIMMER A. (1927) Der soziale Wärmehaushalt der Honigbiene. II Die Wärme der Bienenbrut. Erlanger Jahrbuch für Bienenkunde 5, 1-32.

HIMMER A. (1932) Die Temperaturverhältnisse bei den sozialen Hymenopteren. Biological Reviews 7, 224-253.

HÖLLDOBLER B. (1984) *Evolution of insect communication.* In: Lewis T. (Ed.) Insect Communication, Academic Press, London, United Kingdom.

HÖLLDOBLER B., WILSON E.O. (1990) *The Ants.* Spinger Verlag, Berlin, Germany.

HOFFMAN I. (1966) Gibt es bei Drohnen von *Apis mellifica* L. ein echtes Füttern oder nur eine Futterabgabe? Zeitschrift für Bienenforschung 8, 249-55.

HOFFMANN K. (1978) Thermoregulation bei Insekten. Biologie in unserer Zeit 8, 17-26.

HOGEWEG P., HESPER B. (1983) The ontogeny of the social structure of bumble bee colonies, a MIRROR_model. Behavioral Ecology and Sociobiology 12, 271-283.

HOGEWEG P., HESPER B. (1985) Socioinformatic processes: MIRROR modeling Methodology. Journal of Theoretical Biology 113, 311-330.

References

VAN HONK C.J. G., HOGEWEG P. (1981) The ontogeny of the social structure in a captive *Bombus terrestris* Colony. Behavioral Ecology and Sociobiology 9, 111-119.

HOOVER S.E.R., HIGO H.A., WINSTON M.L. (2006) Worker honeybee ovary development seasonal variation and the influence of larval and adult nutrition. Journal of Comparative Physiology B 176, 55-63.

HUANG Z.-Y., ROBINSON G.E., TOES S.S., YAOI K. J., STRAMBI C., STRAMBI A., STAY B. (1991) Hormonal regulation of behavioural development in the honey bee is based on changes in the rate of juvenile hormone biosynthesis. Journal of Insect Physiology 37, 733-741.

HUANG, Z.-Y., ROBINSON G.E. (1995) Seasonal changes in juvenile hormone in worker honey bees. Journal of Comparative Physiology B 165, 18-28.

HUANG, Z.-Y., ROBINSON G.E. (1996) Regulation of honey bee division of labor by colony age demography. Behavioral Ecology and Sociobiology 39, 147-158.

HUNT J.H. (1991) *Nourishment and the Evolution of the Social Vespidae.* In: The Social Biology of Wasps, Ross K.G., Matthews R.W. (Eds.) Cornell University Press, Ithaka, New York, USA. pp 426-450.

HUNTER J. (1792) Observations on bees. Philosophical Transactions of the Royal Society 82, 128-196.

ISTOMINA-TSVETKOVA K.P. (1960) Contribution to the study trophic relations in adult worker bees, XVII. International Beekeeping Congress Bologna-Roma, Vol. 2, Bucharest, Apimondia, 361-368.

JASSIM O., HUANG Z.-Y., ROBINSON G.E. (2000) Juvenile hormone profiles of worker honey bees, *Apis mellifera*, during normal and accelerated behavioural development. Journal of Insect Physiology 46, 243-249.

JAYCOX E.R., SKOWRONEK W., GWYNN G. (1974) Behavioral changes in worker honey bees (*Apis mellifera*) induced by injections of a juvenile hormone mimic. Annals of the Entomological Society of America 67, 529-534.

JAYCOX E.R. (1976) Behavioral changes in worker honey bees (*Apis mellifera* L.) after injection with synthetic juvenile hormone (Hymenoptera: Apidae). Journal of the Kansas Entomological Society 49, 165-170.

References

JAYCOX E.R., PARISE S.G. (1980) Homesite selection by Italian honey bee swarms, *Apis mellifera ligustica* (Hymenoptera: Apidae). Journal of the Kansas Entomological Society 53, 171-178.

JAYCOX E.R., PARISE S.G. (1981) Homesite selection by swarms of black-bodied honey bees, *Apis mellifera caucasica* and *A. m. carnica* (Hymenoptera: Apidae). Journal of the Kansas Entomological Society 54, 697-703.

JOHANNSMEIER M.F. (Ed.) (2001) Beekeeping in South Africa, 3rd ed., revised, Plant Protection Res. Inst. Handb. No. 14, Agricultural Research Council of South Africa, Pretoria, South Africa.

JOHN M. (1958) Über den Gesamtkohlenhydrat- und Glykogengehalt der Bienen (*Apis mellifica*). Journal of Comparative Physiology A 41, 204-220.

JONGBLOED J., WIERSMA C.A.G. (1934) Der Stoffwechsel der Honigbiene während des Fliegens. Journal of Comparative Physiology A 21, 519-533.

JOSEPHSON R.K. (1981) Temperature and the mechanical performance of insect muscle. In: Heinrich, B. (Ed), Insect Thermoregulation. John Wiley & Sons, New York, USA, 20-24.

KASTBERGER G., STACHL R. (2003) Infrared imaging technology and biological applications. Behavior Research Methods, Instruments, & Computers 35, 429-439.

KEN T., HEPBURN H.R., RADLOFF S.E., YUSHENG Y., YIQIU L., DANYIN Z., NEUMANN P. (2005) Heat-balling wasps by honeybees. Naturwissenschaften 92, 492-495.

KERR W.E. (1969) Some aspects of the evolution of social bees (Apidae). Evolutionary Biology, 3, 119-175.

KISCH J., HAUPT S.S. (2009) Side-specific operant conditioning of antennal movements in the honey bee. Behavioural Brain Research 196, 131-133.

KLEINHENZ M., BUJOK B., FUCHS S., TAUTZ J. (2003) Hot bees in empty broodnest cells: heating from within. Journal of Experimental Biology 206, 4217-4231.

KOEHLER A. (1921) Beobachtungen über Veränderungen am Fettkörper der Biene. Schweizerische Bienen-Zeitung 44, 424-428.

References

KOENIGER N. (1970) Factors determining the laying of drone and worker eggs by the queen honey bee. Bee World 51, 166-169.

KOENIGER N. (1978) Das Wärmen der Brut bei der Honigbiene. Apidologie 9, 305-320.

KORST P.J.A.M., VELTHUIS H.H.W. (1982) The nature of trophallaxis in honeybees. Insectes Sociaux 29, 209-221.

KOVAC H., STABENTHEINER A. (1999) Effect of food quality on the body temperature of wasps (Paravespula vulgaris). Journal of Insect Physiology 45, 183 -190.

KRONENBERG F. (1979) Colonial Thermoregulation in Honey Bees. PhD Dissertation, Stanford University.

KRONENBERG F., HELLER H.C. (1982) Colonial thermoregulation in honey bees (Apis mellifera). Journal of Comparative Physiology B 148, 65-76.

KUBO T., SASAKI M., NAKAMURA J., SASAGAWA H., OHASHI K., TAKEUCHI H. NATORI, S. (1996) Change in the expression of hypopharyngeal-gland proteins of the worker honeybee (Apis mellifera L.) with age and/or role. Journal of Biochemistry 119, 291-295.

KÜHNHOLZ S., SEELEY T.D. (1997) The control of water collection in honey bee colonies. Behavioral Ecology and Sociobiology 41, 407-422.

KUKUK P.F., CROZIER R.H. (1990) Trophallaxis in a communal halictine bee Lasioglossum (Chilalictus) erythrurum. Proceedings of the National Academy of Science 87, 5402-5404.

KUNERT K., CRAILSHEIM K. (1987) Sugar and Protein in the Food for Honeybee Worker Larvae.

In: Chemistry and Biology of Social Insects, Eder J., Rembold H. (Eds.), J Peperny,

München, Germany, pp. 164-165.

KUNIEDA T., FUJIYUKI T., KUCHARSKI R., FORET S., AMENT S.A., TOTH A.L., OHASHI K., TAKEUCHI H., KAMIKOUCHI A., KAGE E., MORIOKA M., BEYE M., KUBO T., ROBINSON G.E., MALESZKA R. (2006) Carbohydrate metabolism genes and pathways in insects: insight from the honey bee genome. Insect Molecular Biology 15, 563-576.

References

KUWABARA M. (1957) Bildung des bedingten Reflexes von Pavlov Typus bei der Honigbiene Apis mellifera. Journal of the Faculty of Science Hokkaido University Zoology 13, 458-464.

VON LACHER V. (1964) Elektrophysiologische Untersuchungen an einzelnen Rezeptoren für Geruch, Kohlendioxid, Luftfeuchtigkeit und Temperatur auf den Antennen der Arbeitsbiene und der Drohne (Apis mellifera L.). Zeitschrift für Vergleichende Physiologie 48, 587-623.

LETZKUS P., RIBI W., WOOD J., ZHU H., ZHANG S., SRINIVASAN,M. (2006) Lateralization of Olfaction in the Honeybee Apis mellifera. Current Biology 16, 1471-1476.

LETZKUS P., BOEDDEKER N., WOOD J.T., ZHANG S., SRINIVASAN,M.V. (2008) Lateralization of visual learning in the honeybee, Biology Letters 4, 16-18.

LIEBIG J.J.H., HÖLLDOBLER B. (1997) Trophallaxis and aggression in the porine ant, Ponera coarctata: implications for the evolution of liquid food exchange in the hymenoptera. Ethology 103, 707-722.

LIN H., WINSTON M.L., HAUNERLAND N.H., SLESSOR K.N. (1999) Influence of age and population size on ovarian development and vitellogenin titers of queenless worker honey bee (Hymenoptera: Apidae). Canadian Entomologist 131, 695-706.

LIN H., DUSSET C., HUANG Z.-Y. (2004) Short-term changes in juvenile hormone titer in honey bee workers due to stress. Apidologie 35, 319-327.

LINDAUER M. (1952) Ein Beitrag zur Frage der Arbeitsteilung im Bienenstaat. Zeitschrift für vergleichende Physiologie 34, 299-345.

LINDAUER M. (1955) The water economy and temperature regulation of the honeybee colony. Bee World 35, 62-72, 81-92 and 105-111.

LOH W., HERAN H. (1970) Wie gut können Bienen Saccharose, Glucose, Fructose und Sorbit im Flugstoffwechsel verwerten? Journal of Comparative Physiology A 67, 436-452.

LOUW G.N., HADLEY N.F. (1985) Water economy of the honeybee: a stoichiometric accounting. The Journal of Experimental Zoology 235, 147-150.

References

MARTIN H., LINDAUER M. (1966) Sinnesphysiologische Leistungen beim Wabenbau der Honigbiene. Zeitschrift für vergleichende Physiologie 53, 372-404.

MAURIZIO A. (1950) Untersuchungen über den Einfluss der Pollennahrung und Brutpflege auf die Lebensdauer und den physiologischen Zustand von Bienen. Schweizerische Bienen-Zeitung 2, 58-64.

MAURIZIO A. (1954) Pollenernährung und Lebensvorgänge bei der Honigbiene (Apis mellifica L.). Landwirtschaftliches Jahrbuch der Schweiz 68, 115-193.

MEYERHOFF G. (1955) „Burgfriede" bei sozialen Insekten, Mitteilungen der Deutschen Entomologischen Gesellschaft 14, 16-17.

MICHENER C.D. (1974) The Social Behavior of the Bees. The Belknap Press of Harvard University Press, Cambridge, Massachusetts, USA.

MICHEU S., CRAILSHEIM K., LEONHARD B. (2000) Importance of proline and other amino acids during honeybee flight. Amino Acids 18, 157-175.

MOCCHEGIANI E., MUZZIOLI M., GIACCONI R. (2000) Zinc and immunoresistance to infection in aging: new biological tools. Trends Pharmacological Science 21, 205-208.

MONTAGNER H., PAIN J. (1971) Étude préliminaire des communications entre ouvrières d'abeilles au cours de la trophallaxie. Insects Sociaux 18, 177-191.

MORITZ R.F.A., HILLESHEIM E. (1985) Inheritance of dominance in honeybees (Apis mellifera capensis Esch.). Behavioral Ecology and Sociobiology 17, 87-89.

MORITZ R.F.A., HALLMEN M. (1986) Trophallaxis of worker honeybees (Apis mellifera L.) of different ages. Insectes Sociaux 33, 26-31.

MORITZ R.F.A., HILLESHEIM E. (1990) Trophallaxis and genetic variance of kin recognition in honeybees, Apis mellifera L. Animal Behavior 40, 641-647.

MORITZ R.F.A., HEISLER T. (1992) Super and half-sister discrimination by honey bee workers (Apis mellifera L.) in a trophallactic bioassay. Insectes Sociaux 39, 365-372.

MORITZ R.F.A., KRYGER P., ALLSOPP M.H. (1996) Competition for royalty in bees. Nature 384, 31.

References

MORITZ R.F.A., SIMON U.E., CREWE R.M. (2000) Pheromonal contest between honeybee workers (*Apis mellifera capensis*). Naturwissenschaften 8, 395-397.

NEUKIRCH A. (1982) Dependence of the life span of the honeybee (*Apis mellifica*) upon flight performance and energy consumption. Journal of Comparative Physiology B 146, 35-40.

NEUMANN P., MORITZ R.F.A., VAN PRAAGH J.P. (1999) Queen mating frequency in different types of honey bee mating apiaries. Journal of Apicultural Research 38, 11-18.

NEUMANN P., MORITZ R.F.A. (2000) Testing genetic variance hypotheses for the evolution of polyandry in the honeybee (*Apis mellifera* L.). Insectes Sociaux 47, 271-279.

NEWSHOLME E. A., CRABTREE B., HIGGINS S. J., THORNTON S. D., START C. (1972) The activities of fructose diphosphatase in flight muscles from the bumble-bee and the role of this enzyme in heat generation. Biochemical Journal 128, 89-97.

NICOLSON S.W., LOUW G.N. (1982) Simultaneous measurement of evaporative water loss, oxygen consumption, and thoracic temperature during flight in a carpenter bee. Journal of Experimental Zoology 222, 287-296.

NICOLSON S.W., HUMAN H. (2008) Bees get a head start on honey production. Biology Letters 4, 299-301.

NIEH J.C., SANCHEZ D. (2005) Effect of food quality, distance and height on thoracic temperature in the stingless bee, *Melipona panamica*. Journal of Experimental Biology 208, 3933-3943.

NIEH J.C., LEÓN A., CAMERON S.A., VANDAME R. (2006) Hot bumble bees at good food: thoracic temperature of feeding *Bombus wilmattae* foragers is tuned to sugar concentration. Journal of Experimental Biology 209, 4185-4191.

NIJHOUT H.F. (1994) *Insect Hormones*. Princeton University Press, Princeton, New Jersey, USA.

NIXON H.L., RIBBANDS C.R. (1952) Food transmission within the honeybee community. Proceedings of the Royal Society of London B 140, 43-50.

NOVAK V. J. A. (1966) *Insect Hormones* (3rd ed.), Methuen, London, United Kingdom.

References

O'DONNELL S. (2001) Worker biting interactions and task performance in a swarm-founding eusocial wasp (*Polybia occidentalis*, Hymenoptera: Vespidae). Behavioral Ecology 12, 353-359.

OHASHI K., SAWATA M., TAKEUCHI H., NATORI S., KUBO T. (1996) Molecular cloning of cDNA and analysis of expression of the gene for α-glucosidase from the hypopharyngeal gland of the honeybee. Biochemical and Biophysical Research Communication 221, 380-385.

OHASHI K., NATORI S., KUBO T. (1997) Change in the mode of gene expression of the hypopharyngeal gland cells with an age-dependent role change of the worker honeybee *Apis mellifera* L. European Journal of Biochemistry 249, 797-802.

OHASHI K., NATORI, S., KUBO, T. (1999) Expression of amylase and glucose oxidase in the hypopharyngeal gland with an age-dependent role change of the worker honeybee (*Apis mellifera* L.). European Journal of Biochemistry 265, 127-133.

OHTANI T. (1974) Behavior repertoire of adult drone honeybee within observation hives. Journal of the Faculty of Science Hokkaido University Zoology 19, 706-721.

OLDROYD B.P., SMOLENSKI A.J., CORNUET J-M., CROZIER R.H. (1994) Anarchy in the beehive. Nature 371, 749.

OLDROYD B.P. (2007) What's Killing American Honey Bees? PLoS Biology 5, e168.

ONO M., OKADA I., SASAKI M. (1987) Heat production by balling in the Japanese honeybee, *Apis cerana japonica* as a defensive behavior against the hornet, *Vespa simillima xanthoptera* (Hymenoptera: Vespidae). Experientia 43, 1031-1032.

ONO M., IGARASHI T., OHNO E., SASAKI M. (1995) Unusual thermal defense by a honeybee against mass attack by hornets. Nature 377, 334-336.

OSTER G.F., WILSON, E.O. (1978) *Caste and Ecology in Social Insects*. Princeton University Press, Princeton, New Jersey, USA.

OTTE D. (1974) Effects and functions in the evolution of signaling systems. Annual Review of Ecolology Evolution and Systematics 5, 385-418.

PAGE R.E (1986) Sperm Utilization in Social Insects. Annual Review of Entomology 31, 297-320.

References

PAGE R.E., LAIDLAW H.H. (1988) Full sisters and super sisters a terminological paradigm. Animal Behavior 36, 944-945.

PANZENBÖCK U., CRAILSHEIM K. (1997) Glycogen in honeybee queens, workers and drones (*Apis mellifera carnica* Pollm.). Journal of Insect Physiology 43, 155-165.

PASCUAL A., HUANG K.L., NEVUE J., PRÉAT T. (2004) Brain asymmetry and long-term memory. Nature 427, 605-606.

PANKIW T., PAGE R.E. (2001) Genotype and colony environment affect honey bee (*Apis mellifera* L.) development and foraging behavior. Behavioral Ecology and Sociobiology **51**, 87-94.

PANZENBÖCK U., CRAILSHEIM K. (1997) Glycogen in honeybee queens, workers and drones (*Apis mellifera carnica*). Journal of Insect Physiology **43**, 155-165.

PAPACHRISTOFOROU A., RORTAIS A., ARNOLD G., IOANNIDES I., SÉRAPHIDES N., GARNERY L., THRASYVOLOU A. (2005) *Defensive behaviour of Apis mellifera cypria against the hornet Vespa orientalis.* Dublin, Ireland.

PAPACHRISTOFOROU A., RORTAIS A., ZAFEIRIDOU G., THEOPHILIDIS G., GARNERY L., THRASYVOULOU A., ARNOLD G. (2007) Smothered to death: hornets asphyxiated by honeybees, Current Biology 17, R795-796.

PARK O.W. (1923) Behavior of water-carriers, American Bee Journal 63, 348-349.

PARK O.W. (1925) The storing and ripening of honey by honeybees. Journal of Economic Entomology 18, 405-410.

PEARCE A.N., HUANG Z.Y., BREED M.D. (2001) Juvenile hormone and aggression in honey bees. Journal of Insect Physiology 47, 1243-1247.

PREEPELOVA L.I., (1929) Biology of laying workers the oviposition of the queen and swarming Bee World 10, 69-71.

PERSHAD S. (1966) L'influence de l'âge sur les échanges de nourriture entre les ouvrières d'abeilles, *Apis mellifica,* Insectes Sociaux 13, 323-328.

PERSHAD S. (1967) Analyze de différents facteurs conditionnant les échanges alimentaires dans une colonie d'abeilles (*Apis mellifera* L.) au moyen du radio-isotope. PhD Dissertation, Toulouse University.

References

PFEIFFER K.J, CRAILSHEIM K. (1997) The influence of drifting on the behaviour of nurse bees Apidologie 28, 198-200.

PIRK C.W.W., NEUMANN P., HEPBURN H.R., MORITZ R.F.A., TAUTZ J. (2004) Egg viability and worker policing in honeybees. Proceedings of the National Academy of Science USA 101, 8649-8651.

POPP J.L., DEVORE I. (1979) *Aggressive competition and social dominance theory: synopsis.* In: Hamburg D.A., McCown E.R. (eds.) The great apes. Cummings, Menlo Park, pp.317-338.

QUELLER D.C., STRASSMANN, J.E. (2002) Kin selection, Current Biology, 12, R832.

RATNIEKS F.L.W (1988) Reproductive harmony via mutual policing by workers in eusocial Hymenoptera. American Naturalist 12, 217-236.

RATNIEKS F.L.W., VISSCHER P.K. (1989) Worker policing in the honeybee. Nature 342, 796-797.

RÉAMUR M.R.A.F. (1742) *Mémoire pour servir à l'histoire des Insectes.* Vol. VI, Imp. Royale

REMBOLD H. (1973) Chemical basis of honeybee caste formation. Proceedings of the VIIth International Congress on Social Insects, London, 327-328.

RITTER W., KOENIGER N. (1977) Influence of the brood on the thermoregulation of honeybee colonies. Proceedings of the 8th Congress of the IUSSI, Wageningen, 283-284.

RITTER W. (1978) Der Einfluß der Brut auf die Änderung der Wärmebildung in Bienenvölkern (*Apis mellifera carnica*). Verhandlungen der Deutschen Zoologischen Gesellschaft, 220.

RITTSCHOF C.C., SEELEY T.D. (2008) The buzz-run: how honeybees signal. 'Time to go!'. Animal Behavior 75, 189-187.

ROBERTS W.C. (1944) Multiple mating of queen bees proved by progeny and flight tests. Gleanings in Bee Culture 72, 255-259.

References

ROBERTSON H.M., WANNER K.W. (2006) The chemoreceptor superfamily in the honey bee, *Apis mellifera*: expansion of the odorant, but not gustatory, receptor family. Genome Research 16, 1395-1403.

ROBINSON G.E., UNDERWOOD B.A., HENDERSON C.E. (1984) A highly specialized water-collecting honey bee. Apidologie 15, 355-358.

ROBINSON G.E. (1985) Effects of a juvenile hormone analogue on honey bee foraging behavior and alarm pheromone production. Journal of Insect Physiology 31, 277-282.

ROBINSON G.E., STRAMBI A., STRAMBI C., PAULINO-SIMÕES Z.L., TOZETO S.O., BARBOSA J.M.N. (1987) Juvenile hormone titers in Africanized and European honey bees in Brazil. General and Comparative Endocrinology 66, 457-459.

ROBINSON G.E., PAGE R.E. (1988) Genetic determination of guarding and undertaking in honey-bee colonies. Nature 333, 356-358.

ROBINSON G.E., PAGE R. (1989) Genetic determination of nectar foraging, pollen foraging, and nest-site scouting in honey bee colonies. Behavioral Ecology and Sociobiology 24, 317-323

ROBINSON G.E., PAGE R., STRAMBI C., STRAMBI A. (1989) Hormonal and genetic control of behavioral integration in honey bee colonies. Science 246, 109-112.

ROBINSON G.E., PAGE R.E., FONDRK M.K. (1990) Intracolonial behavioral variation in worker oviposition, oophagy, and larval care in queenless honey bee colonies. Behavioral Ecology and Sociobiology 26, 315-323.

ROBINSON G.E. (1992) Regulation of division of labor in insect societies. Annual Review of Entomology 37, 637-665.

RÖSCH G.A. (1925) Untersuchungen über die Arbeitsteilung im Bienenstaat. 1. Teil: Die Tätigkeiten im normalen Bienenstaate und ihre Beziehungen zum Alter der Arbeitsbienen. Zeitschrift für vergleichende Physiologie 2, 571-631.

RÖSCH G.A. (1927) Über die Bautätigkeit im Bienenvolk und das Alter der Baubienen. Zeitschrift für vergleichende Physiologie 6, 264-298.

RÖSCH G.A. (1930) Untersuchungen über die Arbeitsteilung im Bienenstaat. Zeitschrift für vergleichende Physiologie 12, 1-71.

References

ROGERS L.J., ANDREW R.J. (eds.) (2002) *Comparative Vertebrate Lateralization.* Cambridge University Press, New York, USA.

ROGERS L.J., SLATER P.J.B., ROSENBLATT J., SNOWDON C., ROPER T. (2002) Lateralization in vertebrates: Its early evolution, general pattern and development. Advances in the Study of Behavior 31, 107-162.

ROGERS L.J., ZUCCA P., VALLORTIGARA G. (2004) Advantage of having a lateralized brain. Proceedings of the Royal Society of London B 271, S420-S422.

ROTH M. (1965) La production de chaleur chez *Apis mellifica* L. Annales de l'Abeille 8, 5-77.

ROTHE U., NACHTIGALL, W. (1989) Flight of the honey bee. IV. Respiratory quotients and metabolic rates during sitting, walking and flying. Journal of Comparative Physiology B 158, 739-749.

ROTHENBUHLER W.C. (1958) Genetics and Breeding of the Honey Bee. Annual Review of Entomology 3, 161-180.

ROTHENBUHLER W.C., KULINCEVIC J.M. KERR W.E. (1968) Bee Genetics. Annual Review of Genetics 2, 413-438.

ROTHENBUHLER W.C., PAGE R.E. (1989) Genetic variability for temporal polyethism in colonies consisting of similar-aged worker honey bees. Apidologie 20, 433-437.

RUTTNER F. (1988) *Biogeography and Taxonomy of Honeybees.* Springer-Verlag. Berlin. Germany.

SACKTOR B. (1970) Regulation of intermediary metabolism, with special reference to the control mechanisms in insect flight muscle. Advances in Insect Physiology 7, 267-347.

SAKAGAMI S.F. (1953) Untersuchungen über die Arbeitsteilung in einem Zwergvolk der Honigbiene. Beiträge zur Biologie des Bienenvolkes, *Apis mellifica* L. Japanese Journal of Zoology 11, 117-185.

SAKAGAMI S.F. (1954) Occurrence of an aggressive behaviour in queenless hives, with considerations on the social organisation of honeybee. Insectes Sociaux 1, 331–343.

References

SASAGAWA H., SAKI M., ACADIA I. (1989) Hormonal control of the division of labor in adult honey bees (*Apis mellifera* L.) I. Effect of methoprene on corpora allata and hypopharyngeal gland, and its α-glucosidase activity. Applied Entomology and Zoology **24**, 66-77.

SCHATTON-GANDELMAYER K., ENGELS W. (1988) Hemolymph proteins and body weight in newly emerged worker honeybees according to the different rates of parasitation by brood mites. Entomologia Generalis **14**, 93-101.

SCHEINER R., PAGE R.E., ERBER J. (2004) Sucrose responsiveness and behavioral plasticity in honey bees (*Apis mellifera*). Apidologie 35, 133-142.

SCHJELDERUP-EBBE T. (1922) Beitrage zur Sozialpsycholgie des Haushuhns. Zeitschrift Psychologie 88, 225-252.

SCHMARANZER S., STABENTHEINER A. (1988) Variability of the Thermal Behavior of Honeybees on a Feeding Place. Journal of Comparative Physiology B 158, 135-141.

SCHMICKL T., CRAILSHEIM K. (2004) Inner nest homeostasis in a changing environment with special emphasis on honey bee brood nursing and pollen supply. Apidologie 35, 249-263.

SCHOLZE E., PICHLER H., HERAN H. (1964) Zur Entfernungsschätzung der Bienen nach dem Kraftaufwand. Naturwissenschaften 51, 69-70.

SEELEY T.D., MORSE R.A. (1976) The nest of the honey bee (*Apis mellifera* L.). Insects Sociaux 23, 495-512.

SEELEY T.D., HEINRICH B. (1981) *Regulation of Temperature in the Nests of Social Insects.* In: Heinrich B. (Ed), Insect Thermoregulation. John Wiley & Sons, Inc., New York, USA, pp. 159-234.

SEELEY T.D. (1982) Adaptive significance of the age polyethism schedule in honeybee colonies. Behavioral Ecolology and Sociobiology 11, 287-293.

SEELEY T.D. (1985) *Honeybee Ecology,* Princeton University Press, Princeton, New Jersey, USA.

SEELEY T.D. (1989) Social foraging in honey bees: how nectar foragers assess their colony´s nutritional status. Behavioral Ecology and Sociobiology 24, 181-199.

References

SEELEY T.D. (1995) *The wisdom of the hive*, Harvard University press, Cambridge Massachusetts.

SIMPSON J. (1961) Nest climate regulation in honey bee colonies: honey bees control their domestic environment by methods based on their habit of clustering together. Science 133, 1327-1333.

SIMPSON J., RIEDEL I.B.M., WILDING M. (1968) Invertase in the hypopharyngeal gland in adult worker honey bees. Apidologie 21, 457-468.

SLEIGH C. (2002) Brave new worlds: trophallaxis and the origin of society in the early twentieth century. Journal of the History of Behavioral Sciences 38, 133-56.

SOLAND-RECKEWEG G., HECKEL G., NEUMANN P., FLURI, P., EXCOFFIER L., (2008) Gene flow in admixed populations and implications for the conservation of the Western honeybee, *Apis mellifera*. Journal of Insect Conservation (in print).

SOTAVALTA O. (1954) On the thoracic temperature of insects in flight. Annals of the Zoological Society 16, 1-22.

SOUTHWICK E.E. (1982) Metabolic energy of intact honey bee colonies. Comparative Biochemistry and Physiology 71 A, 277-281.

SOUTHWICK E.E. (1983) The honey bee cluster as a homeothermic superorganism. Comparative Biochemistry and Physiology 75, 614-645.

SOUTHWICK E.E. (1985) Allometric relations, metabolism and heat conductance in clusters of honey bees at cool temperatures. Journal of Comparative Physiology B 156, 143-149.

SOUTHWICK E.E., HELDMAIER G. (1987) Temperature control in honey bee colonies. BioScience 37, 395-399.

STABENTHEINER A., SCHMARANZER S. (1987) Thermographic determination of body temperatures in honey bees and hornets: calibration and applications. Thermology 2, 563-572.

STABENTHEINER A., SCHMARANZER S. (1988) Flight-related Thermobiological Investigations of Honeybees (*Apis mellifera carnica*). BIONA report 6, 89-102.

References

STABENTHEINER A., HAGMÜLLER K. (1991) Sweet Food Means 'Hot Dancing' in Honey Bees. Naturwissenschaften 78, 471-473.

STABENTHEINER A., KOVAC H., HAGMÜLLER K. (1995) Thermal behavior of round and wagtail dancing honeybees. Journal of Comparative Physiology B 165, 433-444.

STABENTHEINER A. (1996) Effect of foraging, distance on the thermal behaviour of honeybees during walking, dancing and trophallaxis. Ethology 102, 360-370.

STABENTHEINER A., KOVAC H., SCHMARANZER S. (2002) Honeybee nestmate recognition: the thermal behaviour of guards and their examinees. Journal of Experimental Biology 205, 2637-2642.

STABENTHEINER A., PRESSL H., PAPST T., HRASSNIGG N., CRAILSHEIM K. (2003) Endothermic heat production in honeybee winter clusters. The Journal of Experimental Biology 206, 353-358.

STABENTHEINER A., KOVAC H., SCHMARANZER S. (2007) Thermal behaviour of honeybees during aggressive interactions. Ethology 113, 995-1006.

STARKS P.T., GILLEY D.C. (1999) Heat shielding: a novel method of colonial thermoregulation in honey bees. Naturwissenschaften 86, 438-440.

TAUTZ J., MAIER S., GROH, C., RÖSSLER W., BROCKMANN A. (2003) Behavioral performance in adult honey bees is influenced by the temperature experienced during their pupal development. Proceedings of the National Academy of Science 100, 7343-7347.

TAUTZ J., HEILMANN H.R. (2007) *Phänomen Honigbiene*. Spektrum Akademischer Verlag, Heidelberg. Germany.

THERAULAZ G., BONABEAU E., DENEUBOURG J.-L. (1998) The origin of nest complexity in social insects, Complexity 3, 15-25.

TINBERGEN N. (1952) Derived activities: their causation, biological significance, origin and emancipation during evolution. Quarterly Review of Biology 27, 1-32.

UNDERWOOD B.A. (1991) Thermoregulation and energetic decision-making by the honeybees *Apis cerana*, *Apis dorsata* and *Apis laboriosa*. Journal of Experimental Biology 157, 19-34.

References

VALLORTIGARA G., ROGERS L.J. (2005) Survival with an asymmetrical brain: Advantages and disadvantages of cerebral lateralization. Behavioral and Brain Sciences 28, 575-633.

VELTHUIS H.H.W., GERLING D. (1983) At the brink of sociality: interactions between adults of the carpenter bee *Xylocopa pubescens* Spinola. Behavioral Ecology and Sociobiology 12, 209-214.

VILLA J.D., GENTRY C., TAYLOR O.R. JR. (1987) Preliminary observations on thermoregulation, clustering, and energy utilization in African and European honey bees. Journal of the Kansas Entomological Society 60, 4-14.

WADDINGTON K.D. (1990) Foraging profits and thoracic temperature of honey bees (*Apis mellifera*). Journal of Comparative Physiology 160, 325-329.

WCISLO W.T., GONZALEZ V.H. (2006) Social and ecological contexts of trophallaxis in facultatively social sweat bees, *Megalopta genalis* and *M. ecuadoria* (Hymenoptera, Halictidae). Insectes Sociaux 53, 220-225.

WEGENER G. (1996) Flying insects: model systems in exercise physiology. Experientia 52, 404-412.

WEISS K. (1967) Über den Einfluss verschiedenartiger Weiselwiegen auf die Annahme und das Königinnengewicht in der künstlichen Nachschaffungszucht. Zeitschrift für Bienenforschung 9, 121-134.

WEST S.A., GARDNER A., GRIFFIN A.S. (2006) Altruism, Current Biology, 16, R482-R483.

WHEELER W.M. (1928) *The Social Insects: Their Origin and Evolution*. Harcourt, Brace and Co; New York, USA.

WHITE J.W.JR., SUBERS M.H., SCHEPARTZ A.I. (1963) The identification of inhibine, antibacterial factor in the honey, as hydrogen peroxide, and its origin in a honey glucose oxidase. Biochemimica et Biophysica Acta 73, 57-70.

WHITFELD C.W., BEHURA S.K., BERLOCHER S.H. CLARK A.G., JOHNSTON S., SHEPPARD W.S., SMITH D.R., SUAREZ A.V., WEAVER D., NEIL D.T. (2006) Thrice out of Africa: ancient and recent expansion of the honey bee, *Apis mellifera*. Science 314, 642-645.

References

DE WILDE J., BEETSMA J. (1982) The physiology of caste development. Advances in Insect Physiology 16, 167-246.

WILSON E.O. (1971) *The Insects Societies*, Cambridge, Harvard University Press, United Kingdome.

WILSON E.O. (1985A) Between-caste aversion as a basis for division of labor in the ant *Pheidole pubiventris* (Hymenoptera: Formicidae). Behavioral Ecology and Sociobiology 17, 35-37.

WILSON E.O. (1985B) The sociogenesis of insect colonies. Science 228, 1489-1495.

WINSTON M.L., PUNNETT E.N. (1982) Factors determining temporal division of labour in honeybee. Canadian Journal of Zoology 60, 2947-2952.

WINSTON M.L., FERGUSSON L.A. (1985) The effect of worker loss on temporal caste structure in colonies of the honeybee (*Apis mellifera* L.). Canadian Journal of Zoology 63, 777-780.

WINSTON M.L., FERGUSSON L.A. (1986) Influence of the amount of eggs and larvae in honeybee colonies on temporal division of labour. Journal of Apicultural Research 25, 238-241.

WINSTON M.L. (1987) *The Biology of the Honey Bee.* Harvard University Press, Cambridge, Massachusetts. USA.

WOODWORTH C.E. (1936) Effect of reduced temperature and pressure on honeybee respiration. Journal of Economical Entomology 29, 1128-1132.

WOYKE J. (1962) Natural and artificial insemination of queen honeybees. Bee World 43, 21-25.

WOYKE J. (1963) Drone larvae from fertilized eggs of the honeybee. Journal of Apicultural. Research 2, 19 -24.

WYATT G.R., DAVEY K.G. (1996) Cellular and molecular actions of juvenile hormone. II. Roles of juvenile hormone in adult insects. Advances in Insect Physiology 26, 1-155.

Die VDM Verlagsservicegesellschaft sucht für wissenschaftliche Verlage abgeschlossene und herausragende

Dissertationen, Habilitationen, Diplomarbeiten, Master Theses, Magisterarbeiten usw.

für die kostenlose Publikation als Fachbuch.

Sie verfügen über eine Arbeit, die hohen inhaltlichen und formalen Ansprüchen genügt, und haben Interesse an einer honorarvergüteten Publikation?

Dann senden Sie bitte erste Informationen über sich und Ihre Arbeit per Email an *info@vdm-vsg.de*.

Sie erhalten kurzfristig unser Feedback!

VDM Verlagsservicegesellschaft mbH
Dudweiler Landstr. 99   Telefon +49 681 3720 174
D - 66123 Saarbrücken   Fax   +49 681 3720 1749
www.vdm-vsg.de

Die VDM Verlagsservicegesellschaft mbH vertritt

Printed by Books on Demand GmbH, Norderstedt / Germany